PREFACE

The theory of interpolation is a subject which has progressed more slowly than many other branches of mathematics. The reason for this is not far to seek. The practical computer finds himself sufficiently occupied with the carrying out of lengthy calculations, and would rather leave it to the mathematician to provide the necessary tools for these calculations. The mathematician, on the other hand, while undoubtedly interested in the question of expanding functions in interpolation-series, prefers looking at the question from the point of view of expansion in infinite series; but these are of no use to the computer who only wants the first few terms of the series and limits to the error committed. The result is, in many cases, that the class of expansion which is most interesting from the point of view of the mathematician, is the least useful to the computer. Yet a collaboration between the two is necessary in order to obtain satisfactory results.

At present, a non-rigorous point of view is dominant in most text-books on interpolation. Formulas and methods are developed on the assumption that the function under consideration is a polynomial, and thereafter applied to functions which are certainly not polynomials. No distinction is made between the operation of calculating the value of a function previously defined, and that of inserting values in an interval where the function is not defined. Thus the profound difference between a proved fact and an hypothesis is ignored.

Attempts have been made, now and then, to present the subject of interpolation adopting the point of view that only such approximative formulas are to be included for which it is possible to derive a remainder-term simple enough to permit the calculation of limits to the error involved in the formula. The earliest and most conspicuous of these attempts is due to Markoff ("Differenzenrechnung," Leipzig 1896). Without in the least characterizing these attempts as failures, it may be said that the results they contain are too incomplete to enable the computer to do with them alone.

As long as a vast number of the most efficient working formulas were not provided with remainder-terms, it was not to be expected that the new point of view would be welcomed by those for whom it was intended.

The number of formulas with workable remainder-terms has lately increased so much, that although further development is still possible and desirable, a fresh attempt should be made at writing a text-book on the aforesaid lines. This book owes its existence to the lectures the author has in recent years given to actuarial students at the University of Copenhagen, and is, with a few additions and simplifications, a translation of the Danish edition published in 1925. I wish it to be understood that the book is meant as a text-book, and not as a hand-book or encyclopedia on the subject. To carry through with consistency the point of view which appears to me to be the only tenable one, has been my principal aim. References to the literature on the subject have therefore only occasionally been inserted; but the reader who requires a synopsis of what has been written on finite differences, will have an excellent guide in the list of literature appended to Professor N. E. Nörlund's book "Differenzenrechnung" (Berlin, 1924).

The mathematical equipment required in order to master the book is very small. A knowledge of the first principles of the differential and integral calculus should be sufficient. In one or two places a knowledge of the theory of the Gamma-Function has been assumed; but these paragraphs, which have been printed in smaller type, have no organic connection with the rest, and may be left out without damage to the context. The last section of the book, §21, contains some auxiliary theorems of elementary mathematics, and may be read independently of the rest.

The numerical examples have been selected in such a way that they are thought sufficient for illustrating the methods in all cases where the application is not quite obvious. They have not been chosen with a view to taxing the methods to the utmost, but, as should be the case in a text-book, rather with a view to avoiding heavy numerical work which would deter the student from going through the examples by himself. With a table of logarithms, with Barlow's "Table of Squares, etc.," and a product-table such as Crelle's or, still better, Tsuneta Yano's,[1] the reader will be amply equipped for the calculations occurring in the book, although an arithmometer, if at hand, is most useful.

It has, on many occasions, been difficult to find suitable names for the formulas. If the truth be told, most of the formulas of the theory of interpolation should be called "Newton's formula," as appears from "Methodus Differentialis" and other writings of this great mathematician. But practical considerations have taken the place of the historical ones, and I have therefore retained the names by which the formulas usually go (Stirling, Bessel, Cotes, etc.).

I wish to acknowledge my indebtedness to Mr. Arthur Stevenson of the University of Toronto, Canada, who has helped me to polish up the English of my translation. My thanks are also due to The Williams & Wilkins Company for meeting my wishes with respect to the publication of this work.

J. F. Steffensen.
Professor of Actuarial Science
University of Copenhagen

Copenhagen, November, 1925.

[1] Published by: The First Mutual Life Insurance Company, Tokyo, Japan. This table contains in small book-form the same information as Crelle's table.

CONTENTS

		Page
§1.	Introduction	1
§2.	Displacement-Symbols and Differences	4
§3.	Divided Differences	14
§4.	Interpolation-Formulas	22
§5.	Some Applications	34
§6.	Factorial Coefficients	53
§7.	Numerical Differentiation	60
§8.	Construction of Tables	71
§9.	Inverse Interpolation	79
§10.	Elementary Methods of Summation	86
§11.	Repeated Summation	93
§12.	Laplace's and Gauss's Summation-Formulas	104
§13.	Bernoulli's Polynomials	119
§14.	Euler's Summation-Formula	129
§15.	Lubbock's and Woolhouse's Formulas	138
§16.	Mechanical Quadrature	154
§17.	Numerical Integration of Differential Equations	170
§18.	The Calculus of Symbols	178
§19.	Interpolation with Several Variables	203
§20.	Mechanical Cubature	224

Appendix

§21. On Differential Coefficients of Arbitrary Order 231

§1. Introduction

1. The theory of interpolation may, with certain reservations, be said to occupy itself with that kind of information about a function which can be extracted from a *table* of the function. If, for certain values of the variable $x_0, x_1, x_2, \ldots,$ the corresponding values of the function $f(x_0), f(x_1), f(x_2), \ldots$ are known, we may collect our information in the table

$$\begin{array}{cc} x_0 & f(x_0) \\ x_1 & f(x_1) \\ x_2 & f(x_2) \\ \cdot & \cdot \\ \cdot & \cdot \\ \cdot & \cdot \end{array}$$

and the first question that occurs to the mind is, whether it is possible, by means of this table, to calculate, at least approximately, the value of $f(x)$ for an argument not found in the table.

It must be admitted at once, that this problem—"interpolation" in the more restricted sense of the term—cannot be solved by means of these data alone. The table contains no other information than a correspondence between certain numbers, and if we insert another argument x and a perfectly arbitrarily chosen number $f(x)$ as corresponding to that argument, we in no way contradict the information contained in the table.

2. It follows that, if the problem of interpolation is to have a definite, even if approximate, solution, it is absolutely indispensable to possess, beyond the data contained in the table, at least a certain *general idea* of the character of the function. In practice, it is a very general custom to derive formulas of interpolation on the assumption that the function with which we have to deal is a polynomial of a certain degree. The formula will, then, produce exact results if applied to a polynomial of the same or lower degree; but if it is applied to a polynomial of higher degree or to a function which is not a polynomial, nothing whatever is known about the accuracy obtained. In order to justify the application to such cases, it is customary to refer to the fact that most functions can, at least within moderate intervals, be approximated to by polynomials of a suitable degree. But this is only shirking the real difficulty; for if we have to deal with a numerical calculation, it is not sufficient to know that an approximation is obtainable; what we want to know, is *how close* is the approximation actually obtained.

3. A more fertile assumption, the correctness of which can often be ascertained, is that $f(x)$ possesses, in a certain interval, a continuous differential coefficient of a certain order k. It will be shown later, that any such function can be represented as the sum of a polynomial and a "remainder-term" which is of such a simple nature, that if two numbers are known between which the differential coefficient of order k is situated, we can find limits to the error committed by neglecting the remainder-term in the interpolation.

In order to avoid having continuously to revert to the question of the assumptions made about $f(x)$, let it be stated here once and for all, that $f(x)$, where nothing else is expressly said, means a real, single-valued function, continuous in a closed interval, say $a \leq x \leq b$, and possessing in this interval a *continuous* differential coefficient of the highest order of which use is made in deriving each formula under consideration. If this order be k, $f^{(k+1)}(x)$ need not exist at all; much less is there any necessity for assuming that $f(x)$ is a polynomial or at least an analytical function.

Making such liberal assumptions about $f(x)$, we will, nevertheless, be able to solve, in a simple and satisfactory manner, not only the interpolation problem in the restricted sense, but a great number of problems of a similar nature, which, taken together, form the contents of the theory of interpolation in the wider sense of the word.

4. While I assume that the reader is familiar with the notion of a "continuous function" and with the definitions of a differential coefficient and an integral, I make otherwise little use of analysis in this book. Two theorems belonging to elementary mathematical analysis are, however, so frequently referred to, that I find it practical to state them here, together with their proofs:

1. *Rolle's Theorem.* If $f(x)$ is continuous in the closed interval

$a \leqq x \leqq b$ and differentiable in the open interval $a < x < b$, and if $f(a) = f(b) = 0$, it is possible to find at least one point ξ inside the interval, such that

$$f'(\xi) = 0 \qquad (a < \xi < b).$$

Proof. If $f(x)$ is not identically zero in the interval (in which case the theorem is obvious), let m and M be the smallest and largest values it attains. As $f(x)$ is a continuous function, at least one of these values must be attained for some argument ξ situated *between* a and b, as $f(a) = f(b)$. For this value ξ we must have $f'(\xi) = 0$, as otherwise $f(x)$ would be increasing or decreasing in the neighbourhood of ξ and assume values larger and smaller than $f(\xi)$ which is impossible.

2. *The Theorem of Mean Value.* Let $f(x)$ and $\varphi(x)$ be integrable functions of which $f(x)$ is continuous in the closed interval $a \leqq x \leqq b$, while $\varphi(x)$ does not change sign in the interval. There exists, then, at least one point ξ inside the interval such that

$$\int_a^b f(x)\varphi(x)dx = f(\xi)\int_a^b \varphi(x)dx \qquad (a < \xi < b). \tag{1}$$

Proof. Let $\varphi(x)$ be positive (otherwise we may consider $-\varphi(x)$), and let $m \leq f(x) \leq M$. Then

$$m\int_a^b \varphi(x)dx \leq \int_a^b f(x)\varphi(x)dx \leq M\int_a^b \varphi(x)dx.$$

There must, therefore, exist a number μ, intermediate between m and M, such that

$$\int_a^b f(x)\varphi(x)dx = \mu\int_a^b \varphi(x)dx;$$

and as $f(x)$ is continuous, μ may be replaced by $f(\xi)$.

The Theorem of Mean Value is easily extended to double integrals and to sums; we need not go into details.

Occasionally use has been made of results, borrowed from the theory of the Gamma-Function;[1] but the student who is not familiar

[1] See, for instance, Whittaker and Watson: Modern Analysis, third ed., Chapter XII.

with that function can, without any inconvenience, leave out those paragraphs, which have, therefore, been printed in smaller type.

5. Finally, we make considerable use of a simple theorem, belonging to the theory of series. Let us put

$$S = \sum_{0}^{n} U_\nu + R_{n+1}, \tag{2}$$

an equation which serves to define R_{n+1}, if S, U_0, U_1, . . . U_n are given. But we may also look upon (2) as an expansion for S with its remainder-term R_{n+1}. In that case, let the first *non-vanishing* term after U_n be U_{n+s}. We have, then,

$$R_{n+1} = U_{n+s} + R_{n+s+1}. \tag{3}$$

From this is immediately seen, that if R_{n+1} and R_{n+s+1} have opposite signs, then U_{n+s} must have the same sign as R_{n+1} and be numerically larger. This theorem, to which we shall refer as the *Error-Test*, expresses, then, that *if R_{n+1} and R_{n+s+1} have opposite signs, then the remainder-term is numerically smaller than the first rejected, non-vanishing term, and has the same sign.*

It would seem at first, that nothing much is gained by this theorem, as properties of the remainder-term are expressed by other properties belonging to the same; but we shall see later on that the sign of the remainder-term can often be easily determined by means of the known properties of the function to be developed, so that the theorem is of considerable practical value.

It may be noted that it is also seen from (3) that the condition that the two remainder-terms shall have opposite signs, is not only sufficient but also necessary, in order that the theorem shall hold.

§2. Displacement-Symbols and Differences.

6. In the theory of interpolation it is convenient to make use of certain symbols, denoting *operations*. Some of these symbols are important analogues of the symbols, known from the differential and integral calculus, denoting differentiation and integration.

In dealing with finite differences, we must first mention the *symbol of displacement* E^a. If this symbol is prefixed to a function $f(x)$ or, as we shall say for brevity, if E^a is applied to $f(x)$, we mean, that $f(x)$ is changed into $f(x + a)$. We have therefore, according to definition,

$$E^a f(x) = f(x + a).$$

The letter a stands for any real number (positive, zero or negative). We have evidently $E^0 = 1$. If $a = 1$, the exponent is usually left out, so that $E = E^1$.

The practical utility of the displacement-symbol depends on the fact that it obeys certain fundamental laws, and that it is in several respects permissible to operate with it, as if it were a number.

Thus, this symbol possesses the *distributive* property which is expressed in the equation

$$E^a [f(x) + \varphi (x)] = E^a f(x) + E^a \varphi (x).$$

According to this relation, the correctness of which is obvious, the symbol E^a can be applied to the sum of two functions by applying it to each of the functions separately and adding the result.

Two displacement-symbols are said to be "multiplied" with each other, if one is applied after the other to the same function. In this respect, too, these symbols resemble numbers, as the order in which the "factors" are taken is immaterial. It is, for instance, evident that

$$E^a E^b f(x) = E^b E^a f(x).$$

This property is called the *commutative* property; it also holds with respect to a constant k, as

$$E^a k f(x) = k E^a f(x).$$

The "exponents" a, b, etc., resemble real exponents in that

$$E^a E^b f(x) = E^{a+b} f(x);$$

in words, we shall say that the displacement-symbol obeys the *index law*. It follows that the nth power of the symbol denotes the operation *repeated* n times.

The product of two displacement-symbols $(E^a E^b)$ may be looked upon as a single operation E^{a+b}, and it is easily seen that

$$(E^a E^b) E^c f(x) = E^a (E^b E^c) f(x),$$

a relation which expresses the so-called *associative* property, another property which our symbol has in common with numbers.

A *linear function of displacement-symbols*, and consequently any *polynomial* in such symbols, may be considered as an operation, e.g.

$$(k E^a + h E^b) f(x) = k E^a f(x) + h E^b f(x),$$

and this operation has, like the displacement-symbol itself, the distributive, commutative and associative properties. The component parts of the compound operation possess the same properties; thus

$$(k E^a + h E^b) f(x) = (E^b h + E^a k) f(x).$$

The reader will, finally, easily ascertain that two linear functions of displacement-symbols can be multiplied together according to ordinary rules, for instance

$$(k E^a + h E^b)(l E^c + m E^d) f(x)$$
$$= (kl E^{a+c} + hl E^{b+c} + km E^{a+d} + hm E^{b+d}) f(x).$$

It follows from all these properties, taken together, that *polynomials in displacement-symbols can be formed according to the same rules as are valid for numbers.*

On the other hand it is necessary to call attention to the fact, that while division with E^a may without ambiguity be interpreted as multiplication with E^{-a}, it is not yet allowed to divide by a polynomial in displacement-symbols, nor is it permitted to employ infinite series of such symbols. The examination of these questions follows in §18.

In many cases where confusion is not to be feared, the expression of the function $f(x)$ is left out, and the calculation performed with the symbols alone. In that case we write, for instance, $k + E^a$ instead of $k f(x) + f(x + a)$. This often means a considerable economy in writing.

In dealing with functions of two or more variables (or functions) it is often necessary to introduce symbols each acting on one of the variables (or functions) and not on the others. These symbols must then be distinguished from each other, e.g., by denoting them by E_I^a, E_{II}^a, etc. After what has been said, the reader will have no difficulty in satisfying himself that polynomials in E_I^a and E_{II}^a can be formed as if E_I^a and E_{II}^a were numbers.

7. The simplest, and at the same time most important, linear functions of displacement-symbols are the *differences*. In particular, we note the three kinds of differences defined by the equations

$$\left.\begin{aligned}\Delta f(x) &= f(x+1) - f(x)\\ \nabla f(x) &= f(x) - f(x-1)\\ \delta f(x) &= f(x+\tfrac{1}{2}) - f(x-\tfrac{1}{2})\end{aligned}\right\} \quad (1)$$

or, in symbolical form,

$$\left.\begin{aligned}\Delta &= E - 1\\ \nabla &= 1 - E^{-1}\\ \delta &= E^{\frac{1}{2}} - E^{-\frac{1}{2}}.\end{aligned}\right\} \quad (2)$$

They are called respectively the *descending*, the *ascending* and the *central* difference. The reason for this is made clear by a consideration of the following three "difference-tables." In these, the numerical values are the same in corresponding places in all the three tables, so that only the notation differs.

n	$f(n)$	Δ	Δ^2	Δ^3	Δ^4
-2	$f(-2)$				
		$\Delta f(-2)$			
-1	$f(-1)$		$\Delta^2 f(-2)$		
		$\Delta f(-1)$		$\Delta^3 f(-2)$	
0	$f(0)$		$\Delta^2 f(-1)$		$\Delta^4 f(-2)$
		$\Delta f(0)$		$\Delta^3 f(-1)$	
1	$f(1)$		$\Delta^2 f(0)$		
		$\Delta f(1)$			
2	$f(2)$				

n	$f(n)$	∇	∇^2	∇^3	∇^4
-2	$f(-2)$				
		$\nabla f(-1)$			
-1	$f(-1)$		$\nabla^2 f(0)$		
		$\nabla f(0)$		$\nabla^3 f(1)$	
0	$f(0)$		$\nabla^2 f(1)$		$\nabla^4 f(2)$
		$\nabla f(1)$		$\nabla^3 f(2)$	
1	$f(1)$		$\nabla^2 f(2)$		
		$\nabla f(2)$			
2	$f(2)$				

n	$f(n)$	δ	δ^2	δ^3	δ^4
-2	$f(-2)$				
		$\delta f(-\frac{3}{2})$			
-1	$f(-1)$		$\delta^2 f(-1)$		
		$\delta f(-\frac{1}{2})$		$\delta^3 f(-\frac{1}{2})$	
0	$f(0)$		$\delta^2 f(0)$		$\delta^4 f(0)$
		$\delta f(\frac{1}{2})$		$\delta^3(f\frac{1}{2})$	
1	$f(1)$		$\delta^2 f(1)$		
		$\delta f(\frac{3}{2})$			
2	$f(2)$				

It is seen that $f(n)$, $\Delta f(n)$, $\Delta^2 f(n)$, ... for constant n, are always found on the same descending line, $f(n)$, $\nabla f(n)$, $\nabla^2 f(n)$, ... on the same ascending line, while $f(n)$, $\delta^2 f(n)$, $\delta^4 f(n)$ ... are found on a horizontal line, the same applying to $\delta f(n + \frac{1}{2})$, $\delta^3 f(n + \frac{1}{2})$, $\delta^5 f(n + \frac{1}{2})$, ...

8. The three kinds of differences are analogous to the symbol of differentiation D, defined by $Df(x) = f'(x)$. We introduce the three polynomials of degree n

$$\left.\begin{aligned} x^{(n)} &= x(x-1)\ldots(x-n+1) \\ x^{(-n)} &= x(x+1)\ldots(x+n-1) = (-1)^n(-x)^{(n)} \\ x^{[n]} &= x\left(x+\frac{n}{2}-1\right)\left(x+\frac{n}{2}-2\right)\ldots\left(x-\frac{n}{2}+1\right) \\ &= x\left(x+\frac{n}{2}-1\right)^{(n-1)} \end{aligned}\right\} \quad (3)$$

They are called respectively the descending, the ascending and the central *factorial*. For $n = 0$ we assign to them the value 1 (which is also obtained if they are expressed by Gamma-Functions). For $n > 0$ they all contain x as a factor, and therefore vanish for $x = 0$. We now find

$$\begin{aligned} \Delta x^{(n)} &= (x+1)^{(n)} - x^{(n)} \\ &= x^{(n-1)}[(x+1)-(x-n+1)] \\ &= nx^{(n-1)}; \end{aligned}$$

further

$$\begin{aligned} \nabla x^{(-n)} &= x^{(-n)} - (x-1)^{(-n)} \\ &= x^{(-n+1)}[(x+n-1)-(x-1)] \\ &= nx^{(-n+1)}; \end{aligned}$$

finally

$$\delta x^{[n]} = (x+\tfrac{1}{2})\left(x+\frac{n-1}{2}\right)^{(n-1)} - (x-\tfrac{1}{2})\left(x+\frac{n-1}{2}-1\right)^{(n-1)}$$

$$= \left(x+\frac{n-1}{2}-1\right)^{(n-2)}\left[(x+\tfrac{1}{2})\left(x+\frac{n-1}{2}\right)-(x-\tfrac{1}{2})\left(x-\frac{n-1}{2}\right)\right]$$

$$= \left(x + \frac{n-1}{2} - 1\right)^{(n-2)} \cdot nx$$

$$= nx^{[n-1]}.$$

We have, therefore, proved the following important properties:

$$\Delta x^{(n)} = nx^{(n-1)};\ \nabla x^{(-n)} = nx^{(-n+1)};\ \delta x^{[n]} = nx^{[n-1]}; \qquad (4)$$

these relations are quite analogous to the formula $Dx^n = nx^{n-1}$, well known from the differential calculus.

The central factorials of even and odd order may respectively be written

$$\left.\begin{aligned} x^{[2\nu]} &= x^2(x^2-1)(x^2-4)\ldots[x^2-(\nu-1)^2] \\ x^{[2\nu+1]} &= x(x^2-\tfrac{1}{4})(x^2-\tfrac{9}{4})\ldots\left[x^2-\frac{(2\nu-1)^2}{4}\right] \end{aligned}\right\}(5)$$

which shows that $x^{[2\nu]}$ is an even, $x^{[2\nu+1]}$ an odd function of x.

On some occasions the following notation will be found useful:

$$\left.\begin{aligned} x^{[2\nu]-1} &= \frac{x^{[2\nu]}}{x} = x(x^2-1)(x^2-4)\ldots[x^2-(\nu-1)^2] \\ x^{[2\nu+1]-1} &= \frac{x^{[2\nu+1]}}{x} = (x^2-\tfrac{1}{4})(x^2-\tfrac{9}{4})\ldots\left[x^2-\frac{(2\nu-1)^2}{4}\right]. \end{aligned}\right\}(6)$$

The former of these functions is an odd, the latter an even function of x, so that, as in the case of (5) the symbolical exponent is odd when the function is odd, and even when it is even.

9. It is easily seen that Δx^n, ∇x^n and δx^n are all polynomials of degree $n - 1$. Therefore the first difference of a polynomial of degree n will be of degree $n - 1$, the mth of degree $n - m$, the nth a constant, while all the following differences vanish.

It may be concluded from (4), that the nth difference of x^n is $n!$, whether in the case of descending, ascending, or central differences; for x^n may clearly be written as a factorial of degree n *minus* a certain polynomial of degree $n - 1$ which vanishes by repeating the difference-operation n times.

10. On account of the properties we have proved for displacement-symbols we obtain by the binomial theorem

$$\left.\begin{aligned}
\Delta^n f(x) &= (E-1)^n f(x) \\
&= f(x+n) - \binom{n}{1} f(x+n-1) + \binom{n}{2} f(x+n-2) - \ldots + (-1)^n f(x) \\
\nabla^n f(x) &= (1-E^{-1})^n f(x) \\
&= f(x) - \binom{n}{1} f(x-1) + \binom{n}{2} f(x-2) - \ldots + (-1)^n f(x-n) \\
\delta^n f(x) &= (E^{\frac{1}{2}} - E^{-\frac{1}{2}})^n f(x) \\
&= f\left(x+\frac{n}{2}\right) - \binom{n}{1} f\left(x+\frac{n}{2}-1\right) + \ldots + (-1)^n f\left(x-\frac{n}{2}\right).
\end{aligned}\right\} \quad (7)$$

It is seen from these equations, or still more simply by means of the relations $\nabla = E^{-1}\Delta$, $\delta = E^{-\frac{1}{2}}\Delta$, that

$$\nabla^n f(x) = \Delta^n f(x-n), \quad \delta^{2\nu} f(x) = \Delta^{2\nu} f(x-\nu). \quad (8)$$

11. Another important symbol which is a linear function of displacement-symbols is the *mean* $\Box$; defined by[1]

$$\Box\, f(x) = \frac{f(x+\frac{1}{2}) + f(x-\frac{1}{2})}{2}$$

or

$$\Box = \frac{E^{\frac{1}{2}} + E^{-\frac{1}{2}}}{2}. \quad (9)$$

The reader will have no difficulty in proving for himself the following three pairs of formulas of which use will be made later on:

$$\left.\begin{aligned}
\delta x^{[n+1]-1} &= n x^{[n]-1} \\
\Box\, x^{[n+1]-1} &= x^{[n]}.
\end{aligned}\right\} \quad (10)$$

$$\left.\begin{aligned}
\delta^{2\nu} f(0) &= \Delta^{2\nu} f(-\nu) \\
\Box\, \delta^{2\nu+1} f(0) &= \tfrac{1}{2}\left[\Delta^{2\nu+1} f(-\nu) + \Delta^{2\nu+1} f(-\nu-1)\right].
\end{aligned}\right\} \quad (11)$$

$$\left.\begin{aligned}
\delta^{2\nu+1} f(\tfrac{1}{2}) &= \Delta^{2\nu+1} f(-\nu) \\
\Box\, \delta^{2\nu} f(\tfrac{1}{2}) &= \tfrac{1}{2}\left[\Delta^{2\nu} f(-\nu+1) + \Delta^{2\nu} f(-\nu)\right].
\end{aligned}\right\} \quad (12)$$

[1] This symbol is due to Thiele. English authors sometimes write μ, a notation introduced by Sheppard.

Finally, we note the simple relations

$$\square^2 = 1 + \frac{\delta^2}{4}, \qquad E^{\frac{1}{2}} = \square + \frac{\delta}{2}. \tag{13}$$

12. For the nth difference of a product of two functions, a theorem holds which is analogous to Leibnitz' well-known theorem

$$D^n u_x v_x = \sum_{\nu=0}^{n} \binom{n}{\nu} D^{n-\nu} u_x D^{\nu} v_x. \tag{14}$$

Let E_I be a symbol acting on u_x alone, and E_{II} a symbol acting on v_x alone. Then we have

$$\begin{aligned} \Delta u_x v_x &= u_{x+1} v_{x+1} - u_x v_x \\ &= (E_I E_{II} - 1) u_x v_x \end{aligned}$$

or

$$\Delta = E_I E_{II} - 1,$$

so that,[1] as $E_{II} = 1 + \Delta_{II}$,

$$\Delta = E_I \Delta_{II} + \Delta_I \tag{15}$$

and hence

$$\begin{aligned} \Delta^n &= (E_I \Delta_{II} + \Delta_I)^n \\ &= \sum_{\nu=0}^{n} \binom{n}{\nu} \Delta_{II}^{\nu} E_I^{\nu} \Delta_I^{n-\nu} \end{aligned}$$

or, introducing $u_x v_x$,

$$\Delta^n u_x v_x = \sum_{\nu=0}^{n} \binom{n}{\nu} \Delta^{n-\nu} u_{x+\nu} \Delta^{\nu} v_x. \tag{16}$$

This formula which is the required one can, of course, be proved by ordinary induction, but the proof becomes more lengthy.

[1] Formula (15) holds not only for the product of u_x and v_x but for any function of u_x and v_x. The same applies to (17).

Similarly, from

$$\delta = E_I^{\frac{1}{2}}\,\delta_{II} + E_{II}^{-\frac{1}{2}}\,\delta_I \tag{17}$$

we find

$$\delta^n = \sum_{\nu=0}^{n} \binom{n}{\nu} E_I^{\frac{\nu}{2}}\,\delta_{II}^{\nu}\,E_{II}^{-\frac{n-\nu}{2}}\,\delta_I^{n-\nu}$$

and hence

$$\delta^n\,u_x\,v_x = \sum_{\nu=0}^{n} \binom{n}{\nu} \delta^{n-\nu}\,u_{x+\frac{\nu}{2}}\,\delta^{\nu}\,v_{x-\frac{n-\nu}{2}}; \tag{18}$$

a relation which may also be proved by putting, in (16), $\Delta = E^{\frac{1}{2}}\delta$.

13. We have assumed that the table-interval is unity; but a series of corresponding theorems may be established, if the interval is any number $h \equiv \Delta x$. In that case the operation Δ might be defined by

$$\begin{aligned} \Delta f(x) &= f(x+h) - f(x) \\ &= f(x+\Delta x) - f(x). \end{aligned}$$

Writing, under these circumstances,

$$x^{(n)} = x(x-h)\,(x-2h)\;.\;.\;.\;[x-(n-1)h],$$

we find

$$\frac{\Delta\,x^{(n)}}{\Delta\,x} = nx^{(n-1)},$$

tending, for $h \to 0$, to $Dx^n = nx^{n-1}$, while for $h = 1$ the first formula (4) results.

We shall, however, not pursue this extension, but assume henceforth, where nothing else is expressly stated, that the interval is unity; the results for interval h being always easily obtainable by a linear transformation of the variable.

14. Let us now, for a moment, assume that $f(x)$ is a *polynomial* of degree n. It may then be proved that $f(x)$ can be represented by the following four expansions

$$\left.\begin{aligned} f(x) &= \sum_0^n \frac{x^\nu}{\nu!} D^\nu f(0) \\ &= \sum_0^n \frac{x^{(\nu)}}{\nu!} \Delta^\nu f(0) \\ &= \sum_0^n \frac{x^{(-\nu)}}{\nu!} \nabla^\nu f(0) \\ &= \sum_0^n \frac{x^{[\nu]}}{\nu!} \delta^\nu f(0). \end{aligned}\right\} \quad (19)$$

The method of proof is the same in all four cases. If, for instance, we put

$$f(x) = \sum_0^n c_\nu x^{(\nu)},$$

then the coefficients c_ν may be determined by performing on both sides the operation Δ^ν and putting, thereafter, $x = 0$. In this way we find $\Delta^\nu f(0) = \nu!\, c_\nu$ whence $c_\nu = \dfrac{1}{\nu!} \Delta^\nu f(0)$, so that the second expansion (19) has been proved. The other expansions are proved by performing the operations D^ν, ∇^ν and δ^ν on expansions in x^ν, $x^{(-\nu)}$ and $x^{[\nu]}$ respectively with unknown coefficients.

The expansions (19) are all unique. For, if there were two different expansions in, say, $x^{[\nu]}$, their difference would be an expansion of the form

$$0 = k_0 + k_1 x^{[1]} + k_2 x^{[2]} + \ldots + k_n x^{[n]},$$

the coefficients being not all zero. But this is impossible; for k_n must vanish, being the coefficient of x^n, if the polynomial on the right is arranged in ordinary powers of x; for a similar reason k_{n-1} must vanish; and so on.—The expansions in $x^{(\nu)}$ and $x^{(-\nu)}$ may be dealt with in the same way.

15. If $f(x)$ is not a polynomial, expansions in factorials should not be used without examination of the *remainder-term*. Even if such expansions, if continued indefinitely, are convergent and unique—which is not always the case—they are, without remainder-terms, practically useless for interpolation, where they are used for numerical calculation, and where the object is to obtain, by means of the smallest possible number of terms, the desired degree of approximation. In order to derive remainder-terms it is, however, necessary to place the investigation on a broader basis, and for this purpose we make use of the *divided differences* already introduced by Newton[1] which form the subject of the following section.

§3. Divided Differences

16. The 1st, 2nd, . . . r^{th} divided differences of $f(x)$ with respect to the arguments $x_0, x_1, x_2, \ldots$ are defined by the following system of equations

$$\left.\begin{aligned} f(x_0, x_1) &= \frac{f(x_0) - f(x_1)}{x_0 - x_1} \\ f(x_0, x_1, x_2) &= \frac{f(x_0, x_1) - f(x_1, x_2)}{x_0 - x_2} \\ &\vdots \\ f(x_0, \ldots x_r) &= \frac{f(x_0, \ldots x_{r-1}) - f(x_1, \ldots x_r)}{x_0 - x_r} \end{aligned}\right\} \quad (1)$$

where it has, for the time being, been assumed that the arguments $x_0, x_1, \ldots x_r$ are all different. If the values of $f(x)$ employed in these equations are known, they may evidently serve for the successive calculation of the divided differences.

We shall often, where no misunderstanding is to be feared, for brevity denote the rth divided difference by f_r, that is

$$f_r \equiv f(x_0, x_1, \ldots x_r). \quad (2)$$

The difference-table for divided differences has the following appearance

x	$f(x)$	f_1	f_2	f_3
x_0	$f(x_0)$			
		$f(x_0, x_1)$		
x_1	$f(x_1)$		$f(x_0, x_1, x_2)$	
		$f(x_1, x_2)$		$f(x_0, x_1, x_2, x_3)$
x_2	$f(x_2)$		$f(x_1, x_2, x_3)$	
		$f(x_2, x_3)$		
x_3	$f(x_3)$			

[1] See papers by Duncan C. Fraser in the Journal of the Institute of Actuaries, vol. LI, p. 77 and p. 211.

It is, however, clear that it is possible, with the given values of $f(x)$, to form several more divided differences than those stated in the table, for instance $f(x_0, x_2)$, $f(x_0, x_1, x_3)$, etc. But for obvious reasons we content ourselves with a number of differences, sufficient for interpolation-purposes, and when the succession of arguments $x_0, x_1, x_2, \ldots$ (which need not be arranged in increasing order of magnitude) has once been settled, the difference-schedule has the appearance shown above.

17. It is clear that f_r must be a linear function of $f(x_0)$, $f(x_1)$, $\ldots . f(x_r)$, and we may without difficulty derive an explicit expression for this function. We write, for abbreviation,

$$\left.\begin{aligned} P(x) &= (x - x_0)(x - x_1) \ldots (x - x_r) \\ P_\nu(x) &= \frac{P(x)}{x - x_\nu} \end{aligned}\right\} \quad (3)$$

and prove by induction that

$$f_r = \sum_0^r \frac{f(x_\nu)}{P_\nu(x_\nu)}. \qquad (4)$$

For this formula is valid for $r = 1$ in which case it reduces to

$$f_1 = \frac{f(x_0)}{x_0 - x_1} + \frac{f(x_1)}{x_1 - x_0} = \frac{f(x_0) - f(x_1)}{x_0 - x_1}.$$

Let us now assume that it has been proved for $f_1, f_2, \ldots f_r$; it may, then, be proved that it is also valid for f_{r+1}; for by (4) and the last equation (1) we find

$$f_{r+1} = \left[\Sigma \frac{f(x_\nu)}{(x_\nu - x_0)\ldots(x_\nu - x_r)} - \Sigma \frac{f(x_\nu)}{(x_\nu - x_1)\ldots(x_\nu - x_{r+1})}\right] \frac{1}{x_0 - x_{r+1}}$$

$$= \Sigma \frac{f(x_\nu)}{(x_\nu - x_0)\ldots(x_\nu - x_{r+1})},$$

it being understood that the factor $(x_\nu - x_\nu)$ must be left out in all the denominators. But this expression is identical with (4), if r is replaced by $r + 1$ in the latter equation.

From (4) we obtain the important theorem, that f_r *is symmetrical in all the* $r+1$ *arguments*, so that the order in which these are taken is indifferent, e.g.

$$f(x_0, x_1, \ldots x_r) = f(x_r, x_0, \ldots x_{r-1}).$$

18. We may express f_r in several other ways.[1] Let θ_p be an operation which, applied to $f(x_0)$, changes this function into

$$\theta_p f(x_0) = \frac{f(x_0) - f(x_p)}{x_0 - x_p} = f(x_0, x_p). \tag{5}$$

The symbol θ_p acts on x_0 alone. It has the *distributive* property, as

$$\theta_p [f(x_0) + \varphi(x_0)] = \theta_p f(x_0) + \theta_p \varphi(x_0),$$

and the operations θ_p and θ_q are *commutative*, for

$$\theta_p \theta_q f(x_0) = \theta_p \frac{f(x_0) - f(x_q)}{x_0 - x_q}$$

$$= \frac{f(x_0)}{(x_0 - x_p)(x_0 - x_q)} + \frac{f(x_p)}{(x_0 - x_p)(x_q - x_p)} + \frac{f(x_q)}{(x_0 - x_q)(x_p - x_q)}$$

and this expression is symmetrical in p and q.

It is now easily proved by induction that

$$f_r = \theta_r \theta_{r-1} \ldots \theta_1 f(x_0). \tag{6}$$

For this theorem is obvious for $r = 1$, and if it has been proved as far as r inclusive, we find

$$\theta_{r+1} \theta_r \ldots \theta_1 f(x_0) = \theta_{r+1} f(x_0, x_1, \ldots x_r)$$

$$= \frac{f(x_0, x_1, \ldots x_r) - f(x_{r+1}, x_1, \ldots x_r)}{x_0 - x_{r+1}}$$

$$= f_{r+1},$$

as $f(x_{r+1}, x_1, \ldots x_r) = f(x_1, x_2, \ldots x_{r+1})$, f_r being a symmetrical function of all the arguments.

[1] J. L. W. V. Jensen: Sur une expression simple du reste dans la formule d'interpolation de Newton. Bulletin de l'Académie Royale de Danemark, 1894, p. 246.

19. A representation of f_r by means of integrals is obtained as follows. It is found directly, by performing the integration, that

$$\theta_p f(x_0) = \int_0^1 f'\,[tx_0 + (1-t)x_p]\,dt.$$

The operation θ_p may, therefore, be performed in 3 successive steps:

1. Differentiating with respect to x_0.
2. Replacing x_0 by $tx_0 + (1-t)\,x_p$.
3. Integrating the result with respect to t from 0 to 1.

We have now, to begin with, interpreting t_0 as 1,

$$f_1 = \theta_1 f(x_0) = \int_0^1 f'\,(x_0t_1 - x_1 \nabla t_1)\,dt_1,$$

and thereafter, by induction,

$$f_r = \int_0^1 dt_1 \int_0^{t_1} dt_2 \ldots \int_0^{t_{r-1}} dt_r\, f^{(r)}\left(x_0t_r - \sum_1^r x_\nu \nabla t_\nu\right);$$

for, performing on both sides the operation θ_{r+1} and replacing thereafter t_rt by t_{r+1}, we obtain an equation of the same form where r has been replaced by $r+1$. For the convenience of the reader we will give the proof in detail, using for abbreviation the notation

$$\int^{(r)} \equiv \int_0^1 dt_1 \int_0^{t_1} dt_2 \ldots \int_0^{t_{r-1}} dt_r, \tag{7}$$

so that the relation to be proved may be written

$$f_r = \int^{(r)} f^{(r)}\left(x_0\,t_r - \sum_1^r x_\nu \nabla t_\nu\right).$$

We now proceed by the three steps mentioned above.

1. Differentiating with respect to x_0, we get

$$\int^{(r)} t_r f^{(r+1)}\left(x_0\,t_r - \sum_1^r x_\nu \nabla\cdot t_\nu\right).$$

2. Replacing now x_0 by $tx_0 + (1-t)x_{r+1}$, we find

$$\int^{(r)} t_r f^{(r+1)}\left(t_rt(x_0 - x_{r+1}) + t_r\,x_{r+1} - \sum_1^r x_\nu \nabla t_\nu\right).$$

3. Integrating this with respect to t from 0 to 1, we introduce the new variable $t_{r+1} = t_r t$, so that $dt_{r+1} = t_r\, dt$. The result may be written

$$\int^{(r)} \int_0^{t_r} dt_{r+1} f^{(r+1)} \left(x_0 t_{r+1} - \sum_1^{r+1} x_\nu \nabla t_\nu \right)$$
$$= \int^{(r+1)} f^{(r+1)} \left(x_0 t_{r+1} - \sum_1^{r+1} x_\nu \nabla t_\nu \right);$$

but this is the same result as that which is obtained by replacing r by $r + 1$ in the above expression for f_r.

This expression which has thus been proved may be written more neatly by considering that f_r is symmetrical in the arguments $x_0, x_1, \ldots x_r$, so that instead of $x_0 t_r - \sum_1^r x_\nu \nabla t_\nu$ we may write

$$x_r t_r - \sum_1^r x_{\nu-1} \nabla t_\nu = x_0 + \sum_1^r t_\nu \nabla x_\nu.$$

We thus obtain the formula

$$f_r = \int^{(r)} f^{(r)} \left(x_0 + \sum_1^r t_\nu \nabla x_\nu \right). \tag{8}$$

We shall refer to this formula as Jensen's formula, although it has been found independently by several authors.[1]

As $0 \leq t_\nu \leq 1$, the argument $x_0 + \sum_1^r t_\nu \nabla x_\nu$ is comprised between the smallest and the largest of the numbers $x_0, x_1, \ldots x_r$, if these are arranged in increasing order of magnitude.

A limiting case of particular interest arises if we put $x_0 = x$; $x_1 = x_2 = \ldots = x_r = 0$; $r = n + 1$. In that case we find, as (8) retains a meaning for coinciding arguments,

$$f_{n+1} = \int^{(n+1)} f^{(n+1)} (x - t_1 x)$$
$$= \frac{1}{n!} \int_0^1 t_1^n f^{(n+1)} (x - t_1 x)\, dt_1,$$

[1] See Runge and Willers: Numerische und graphische Integration, p. 67 (Encyklopädie der mathematischen Wissenschaften, Bd. II, 3). Nörlund: Differenzenrechnung p. 16.

or, if we put $t_1 = 1 - t$,

$$f_{n+1} = \frac{1}{n!}\int_0^1 (1-t)^n f^{(n+1)}(xt)\,dt. \tag{9}$$

20. The divided differences possess a number of interesting properties, amongst which we note the following ones:[1]

1. It follows from the distributive and commutative properties of the operation θ_p that if $f(x) = k\,\varphi(x) + h\,\psi(x)$, then $f_r = k\varphi_r + h\psi_r$.

2. A change of origin without change of the tabular values of $f(x)$ does not influence the divided differences, as these only depend on the given values of the function and on the *differences* between arguments. If, on the other hand, the argument is multiplied by a constant factor, while the tabular values of $f(x)$ are left unchanged, then f_r must be divided by the rth power of this constant.

3. If $f(x_0)$ is a polynomial of the nth degree in x_0, then f_r is a polynomial of the $(n-r)^{th}$ degree in x_0. For the operation θ_p reduces the degree by unity, as is seen from (5) where the numerator must contain the factor $x_0 - x_p$, as it vanishes for $x_0 = x_p$.

4. If $f(x) = x^n$, then

$$f_r = \Sigma\, x_0^{\alpha_0} x_1^{\alpha_1} \ldots x_r^{\alpha_r} \quad (\alpha_0 + \alpha_1 + \ldots + \alpha_r = n - r), \tag{10}$$

the summation being extended to all the different products where the sum of the exponents is $n - r$.

The theorem is easily proved by induction, as

$$f_1 = \theta_1 x_0^n = \frac{x_0^n - x_1^n}{x_0 - x_1}$$

$$= x_0^{n-1} + x_0^{n-2}x_1 + \ldots + x_1^{n-1},$$

and $\theta_{r+1} f_r = f_{r+1}$. In particular we have

$$f_{n-1} = \sum_0^{n-1} x_\nu,\ f_n = 1,\ f_{n+\nu} = 0\ (\nu > 0). \tag{11}$$

The expression (10) is the coefficient of t^{n-r} in the development of

$$\frac{1}{(1 - tx_0)(1 - tx_1)\ldots(1 - tx_r)} \tag{12}$$

[1] Thiele: Interpolationsrechnung, §5, §7.

or

$$(1 + tx_0 + t^2x_0^2 + \ldots)(1 + tx_1 + t^2x_1^2 + \ldots)\ldots(1 + tx_r + t^2x_r^2 + \ldots)$$

in powers of t, taking t^{α_0} from the first, t^{α_1} from the second, ... t^{α_r} from the last bracket, and making $\sum_0^r \alpha_\nu = n - r$. The number of terms in (10) is obtained by putting $x_0 = x_1 = \ldots = x_r = 1$; it therefore equals, according to (12), the coefficient of t^{n-r} in the expansion of $(1 - t)^{-r-1}$, that is $\binom{n}{r}$.

5. If $f(x) = \frac{1}{x}$, we have

$$f_r = \frac{(-1)^r}{x_0 x_1 \cdots x_r}. \tag{13}$$

For we find in succession

$$\theta_1 \frac{1}{x_0} = \frac{\frac{1}{x_0} - \frac{1}{x_1}}{x_0 - x_1} = -\frac{1}{x_0 x_1},$$

$$\theta_2 \theta_1 \frac{1}{x_0} = \frac{1}{x_0 x_1 x_2}, \text{ etc.}$$

6. It appears from (1) that if $x_\nu = \nu$, then $f_r = \frac{\Delta^r f(0)}{r!}$. The rth divided difference is, therefore, not comparable with the rth difference but is generally, numerically speaking, considerably smaller.

7. If all the arguments coincide, we have

$$f_r = \frac{f^{(r)}}{r!}, \tag{14}$$

as can be seen from (9) for $x = 0$.

Cases where some, but not all, of the arguments coincide, are treated by (1), avoiding the divisor 0 and making use of (14).

We find, for instance,

$$f(a, a, b) = \frac{f(a, a) - f(a, b)}{a - b} = \frac{f'(a) - f(a, b)}{a - b},$$

$$f(a, a, a, b) = \frac{f(a, a, a) - f(a, a, b)}{a - b}$$

$$= \tfrac{1}{2} \frac{f''(a)}{a - b} - \frac{f'(a) - f(a, b)}{(a - b)^2}.$$

In practical operations with such divided differences with *repeated arguments*, the difference-table has, for instance, the following appearance

x	$f(x)$	f_1	f_2	f_3
a	$f(a)$			
		$f(a, a)$		
a	$f(a)$		$f(a, a, a)$	
		$f(a, a)$		$f(a, a, a, b)$
a	$f(a)$		$f(a, a, b)$	
		$f(a, b)$		
b	$f(b)$			

If, on the other hand, we had written $f(a, b, a)$ and $f(a, a, b, a)$ instead of $f(a, a, b)$ and $f(a, a, a, b)$, the method employed above would have resulted in the indeterminate forms

$$f(a, b, a) = \frac{f(a, b) - f(b, a)}{a - a} = \frac{0}{0},$$

$$f(a, a, b, a) = \frac{f(a, a, b) - f(a, b, a)}{a - a} = \frac{0}{0}.$$

The indetermination can, of course, be avoided by writing a, $a + \epsilon$, $a + 2\epsilon$, ... instead of a, a, a, ... and letting $\epsilon \to 0$; in this way we arrive at the same results as before, but by detour.

21. If $f(x)$ is a polynomial of degree n, $f(x, a_0, a_1, \ldots a_n)$ vanishes, according to No. 20, 3°, as this expression is a divided difference of order $n + 1$. We therefore have, by (4),

$$0 = \frac{f(x)}{(x - a_0)(x - a_1) \ldots (x - a_n)} + \frac{f(a_0)}{(a_0 - x)(a_0 - a_1) \ldots (a_0 - a_n)}$$

$$+ \ldots + \frac{f(a_n)}{(a_n - x)(a_n - a_0) \ldots (a_n - a_{n-1})}.$$

If, by analogy with (3), we put

$$\left.\begin{aligned} P(x) &= (x - a_0)(x - a_1) \ldots (x - a_n) \\ P_\nu(x) &= \frac{P(x)}{x - a_\nu} \end{aligned}\right\} \quad (15)$$

the formula may be written

$$f(x) = \sum_{\nu=0}^{n} \frac{P_\nu(x)}{P_\nu(a_\nu)} f(a_\nu). \qquad (16)$$

This is Lagrange's formula. It is seen directly that for $x = a_\nu$ the right-hand side assumes the value $f(a_\nu)$. The formula may, therefore, be used for finding the polynomial of degree n which, for $n + 1$ given values of the variable, assumes given values. It is well known that only one such polynomial exists.

§4. Interpolation-Formulas

22. If in §3 (1) we replace $x_0, x_1, \ldots x_r$ by $x, a_0, a_1, \ldots a_n$, this system of equations may be written

$$\left.\begin{aligned} f(x) &= f(a_0) + (x - a_0) f(x, a_0) \\ f(x, a_0) &= f(a_0, a_1) + (x - a_1) f(x, a_0, a_1) \\ &\ldots\ldots\ldots\ldots\ldots\ldots \\ f(x, a_0, \ldots a_{n-1}) &= f(a_0, \ldots a_n) + (x - a_n) f(x, a_0, \ldots a_n) \end{aligned}\right\}$$

From this, we obtain in succession

$$\begin{aligned} f(x) &= f(a_0) + (x - a_0) f(x, a_0) \\ &= f(a_0) + (x - a_0) f(a_0, a_1) + (x - a_0)(x - a_1) f(x, a_0, a_1) \\ &= \ldots\ldots\ldots\ldots\ldots\ldots \end{aligned}$$

and, in the general case,

$$\left.\begin{aligned} f(x) = f(a_0) &+ (x - a_0) f(a_0, a_1) + (x - a_0)(x - a_1) f(a_0, a_1, a_2) + \ldots \\ &+ (x - a_0) \ldots (x - a_{n-1}) f(a_0, \ldots a_n) + R \end{aligned}\right\} \quad (1)$$

where

$$R = (x - a_0) \ldots (x - a_n) f(x, a_0, \ldots a_n). \qquad (2)$$

Formula (1) is due to Newton and is called *Newton's interpolation-formula with divided differences.*

The remainder-term (2) may, by means of §3(8) and §3(15) be written in the more useful form

$$R = P(x) \int^{(n+1)} f^{(n+1)} \left[(1 - t_1)\, x + \sum_{1}^{n+1} t_\nu \nabla a_{\nu-1} \right], \tag{3}$$

putting $a_{-1} = 0$.

If we apply the Theorem of Mean Value to the integral, $f^{(n+1)}$ being the function f of §1(1) while φ is a constant, we obtain the practical form of the remainder-term

$$R = P(x) \frac{f^{(n+1)}(\xi)}{(n+1)!}, \tag{4}$$

ξ being situated between the smallest and the largest of the numbers x, a_0, a_1, . . . a_n, as appears from the observations made in No. 19 concerning §3(8).

The form (4) for the remainder-term lends itself easily to the application of the Error-Test established under No. 5. For if $f^{(n+1)}$ and $f^{(n+2)}$ both have constant signs, we may always choose a_{n+1} on such a side of x that the two remainder-terms

$$R_{n+1} = (x - a_0) \ldots (x - a_n)\, f(x, a_0, \ldots a_n)$$

and

$$R_{n+2} = (x - a_0) \ldots (x - a_{n+1})\, f(x, a_0, \ldots a_{n+1})$$

have opposite signs.

If we only know that $f^{(n+2)}$ has constant sign, we may proceed as follows.[1] Let us assume that x is comprised between the limits l and L, and that $f^{(n+2)}$ does not change its sign within an interval comprising all the numbers l, L, a_0, a_1, . . . a_n. Then $f(x, a_0, a_1, \ldots a_n)$, if not constant, is either constantly increasing or constantly decreasing in the interval and is, therefore, comprised between the limits $f(l, a_0, a_1, \ldots a_n)$ and $f(L, a_0, a_1, \ldots a_n)$.

[1] Inge Lehmann: On the Accuracy of Interpolation. Transactions of the 6th Scandinavian Mathematical Congress, Copenhagen, 1926, p. 375.

23. It is also possible to derive (4) without making use of Jensen's formula. Let us for a moment write (1) in the form

$$f(x) = Q(x) + R(x), \tag{5}$$

$Q(x)$ being the polynomial of degree n, $R(x)$ the remainder-term. As it has been assumed (see No. 3) that $f^{(n+1)}(x)$ is a continuous function, $f(x, a_0, a_1, \ldots a_n)$ must be finite, even if some of the arguments coincide. It is, then, seen from (2) that $R(x)$ vanishes at the $n + 1$ points $a_0, a_1, \ldots a_n$ which we assume, for the moment, to be all different. But according to Rolle's theorem the differential coefficient $R'(x)$ must, then, vanish at least n times, $R''(x)$ consequently at least $n - 1$ times, etc., and finally $R^{(n)}(x)$ at least once in every interval comprising the arguments $a_0, a_1, \ldots a_n$. Now, let ξ be such a number that $R^{(n)}(\xi) = 0$. We find, then, by (5),

$$R^{(n)}(\xi) = f^{(n)}(\xi) - Q^{(n)}(\xi) = 0;$$

but $Q(x)$ being of degree n, $Q^{(n)}(x)$ must be a constant, and it is seen immediately that the value of this constant is $n!\, f(a_0, a_1, \ldots a_n)$. We have thus proved the important formula

$$f(a_0, a_1, \ldots a_n) = \frac{f^{(n)}(\xi)}{n!} \tag{6}$$

which evidently still subsists, if some of the arguments coincide.

If, in (6), we introduce another argument x, we have

$$f(x, a_0, \ldots a_n) = \frac{f^{(n+1)}(\xi)}{(n+1)!},$$

the new ξ being comprised between the smallest and the largest of the numbers $x, a_0, a_1, \ldots a_n$. But if we insert this expression in (2), we have again (4).

24. If, in particular, $f(x)$ is a polynomial of degree n, $f(x, a_0, a_1, \ldots a_n)$ vanishes, so that (1) reduces to

$$\left.\begin{aligned} f(x) = f(a_0) + (x-a_0)\, f(a_0, a_1) + (x-a_0)(x-a_1)\, f(a_0, a_1, a_2) + \ldots \\ + (x - a_0) \ldots (x - a_{n-1})\, f(a_0, a_1, \ldots a_n). \end{aligned}\right\} \tag{7}$$

This formula may evidently be used for determining the polynomial of degree n, for which $f(a_0)$, $f(a_0, a_1)$, . . . $f(a_0, \ldots a_n)$ assume given values. Only one expansion of this nature exists which may be proved in the same way in which we proved that the expansions §2(19) are unique. As both (7) and Lagrange's formula §3(16), express that $f(x, a_0, a_1, \ldots a_n)$ vanishes if $f(x)$ is a polynomial of degree n, these formulas must be identical. Lagrange's formula may, therefore, be used for interpolation to exactly the same extent as Newton's formula with divided differences and *has the same remainder-term.* For actual numerical interpolation Lagrange's formula is, however, as a rule not very suitable.

25. It should be noted that (1) does not lose in generality by putting $x = 0$, provided that all the numbers a_ν are arbitrary; for this only means that the origin is transferred to x. We thus obtain the simplified form

$$\left.\begin{aligned} f(0) &= f(a_0) - a_0 f_1 + a_0 a_1 f_2 - \ldots + (-1)^n a_0 \ldots a_{n-1} f_n + R, \\ R &= (-1)^{n+1} a_0 \ldots a_n \frac{f^{(n+1)}(\xi)}{(n+1)!}. \end{aligned}\right\} \quad (8)$$

26. We may now write down some important particular cases of the interpolation-formula with divided differences. As regards the notation, ξ always means a number comprised between the smallest and the largest of the arguments employed, R the remainder-term; it must, therefore, not be expected that two numbers which are both denoted by ξ or by R are identical.

If, in (1) and (4) we put in succession $a_\nu = 0$, $a_\nu = \nu$ and $a_\nu = -\nu$, we obtain the three formulas

$$f(x) = \sum_0^n \frac{x^\nu}{\nu!} f^{(\nu)}(0) + \frac{x^{n+1}}{(n+1)!} f^{(n+1)}(\xi), \quad (9)$$

$$f(x) = \sum_0^n \frac{x^{(\nu)}}{\nu!} \Delta^\nu f(0) + \frac{x^{(n+1)}}{(n+1)!} f^{(n+1)}(\xi), \quad (10)$$

$$f(x) = \sum_0^n \frac{x^{(-\nu)}}{\nu!} \nabla^\nu f(0) + \frac{x^{(-n-1)}}{(n+1)!} f^{(n+1)}(\xi). \quad (11)$$

In (9) ξ is comprised between 0 and x, in (10) between the smallest and the largest of the numbers 0, n, x, in (11) between the smallest and the largest of the numbers $0, -n, x$. The first of these formulas is Maclaurin's expansion, the second the interpolation formula with descending differences, the third the corresponding formula with ascending differences. If $f(x)$ is a polynomial of degree n, the three formulas are identical with the corresponding developments in §2(19), as in that case the remainder-terms vanish. But (9), (10) and (11) have a far wider field of application, being valid also when $f(x)$ is not a polynomial, the only assumption being, that $f^{(n+1)}(x)$ exists and is continuous. If this differential coefficient can be enclosed between known and sufficiently narrow limits, these formulas may therefore, as well as (1) itself, be employed for the approximate calculation of $f(x)$ for an argument not contained in the difference-table, and that in such a way that limits can be assigned to the error committed.

It should still be noted that the remainder-term in (9) may be written in the well-known form

$$R = \frac{x^{n+1}}{n!}\int_0^1 (1-t)^n f^{(n+1)}(xt)\,dt, \tag{12}$$

as follows from §3(9).

27. A class of interpolation-formulas, called the Gauss'ian formulas, is obtained by putting, in (1), $a_{2\nu} = \nu$, $a_{2\nu-1} = -\nu$. We obtain, stopping at the order $2k$ and putting $\frac{1}{(-1)!} = 0$,

$$\left.\begin{aligned} f(x) &= \sum_0^k \left[\frac{x^{[2\nu]-1}}{(2\nu-1)!}\Delta^{2\nu-1}f(-\nu) + \frac{(x+\nu)\,x^{[2\nu]-1}}{(2\nu)!}\Delta^{2\nu}f(-\nu)\right] + R \\ R &= \frac{x^{[2k+2]-1}}{(2k+1)!}f^{(2k+1)}(\xi), \end{aligned}\right\} \tag{13}$$

while, stopping at the order $2k-1$, we find

$$\left.\begin{aligned} f(x) &= \sum_0^{k-1}\left[\frac{(x+\nu)\,x^{[2\nu]-1}}{(2\nu)!}\Delta^{2\nu}f(-\nu) + \frac{x^{[2\nu+2]-1}}{(2\nu+1)!}\Delta^{2\nu+1}f(-\nu-1)\right] + R \\ R &= \frac{(x+k)\,x^{[2k]-1}}{(2k)!}f^{(2k)}(\xi). \end{aligned}\right\} \tag{14}$$

If, on the other hand, we put $a_{2\nu} = -\nu$, $a_{2\nu-1} = \nu$ and stop at the order $2k$, we have

$$\left.\begin{aligned} f(x) &= \sum_0^k \left[\frac{x^{[2\nu]-1}}{(2\nu-1)!}\Delta^{2\nu-1}f(-\nu+1) + \frac{(x-\nu)\,x^{[2\nu]-1}}{(2\nu)!}\Delta^{2\nu}f(-\nu)\right] + R \\ R &= \frac{x^{[2k+2]-1}}{(2k+1)!}f^{(2k+1)}(\xi), \end{aligned}\right\} \quad (15)$$

and, stopping at the order $2k - 1$,

$$\left.\begin{aligned} f(x) &= \sum_0^{k-1} \left[\frac{(x-\nu)\,x^{[2\nu]-1}}{(2\nu)!}\Delta^{2\nu}f(-\nu) + \frac{x^{[2\nu+2]-1}}{(2\nu+1)!}\Delta^{2\nu+1}f(-\nu)\right] + R \\ R &= \frac{(x-k)\,x^{[2k]-1}}{(2k)!}f^{(2k)}(\xi). \end{aligned}\right\} \quad (16)$$

28. A formula often used in practice is obtained by forming the arithmetical mean of (13) and (15). We find, by §2 (11),

$$\left.\begin{aligned} f(x) &= \sum_0^k \left[\frac{x^{[2\nu]-1}}{(2\nu-1)!}\,\square\,\delta^{2\nu-1}f(0) + \frac{x^{[2\nu]}}{(2\nu)!}\delta^{2\nu}f(0)\right] + R \\ R &= \frac{x^{[2k+2]-1}}{(2k+1)!}f^{(2k+1)}(\xi). \end{aligned}\right\} \quad (17)$$

This formula is called Stirling's interpolation-formula, although it was already found by Newton.

If, in (17), we stop at the order $2k - 1$, the remainder-term does not assume a similar simple form, but we may at any rate put

$$R = \frac{x^{[2k]}}{(2k)!}f^{(2k)}(\xi_1) + \frac{x^{[2k+2]-1}}{(2k+1)!}f^{(2k+1)}(\xi_2), \quad (18)$$

as $\delta^{2k}f(0) = f^{(2k)}(\xi_1)$.

In (17) we have employed differences which in the difference-table are placed on the horizontal line starting at $f(0)$, together with arithmetical means of the differences placed just above and below that line.

It is worth noting that Stirling's formula may also be derived from (13) alone, if the expression in brackets is written in the form

$$\frac{x^{[2\nu]-1}}{(2\nu-1)!}[\Delta^{2\nu-1}f(-\nu)+\tfrac{1}{2}\Delta^{2\nu}f(-\nu)]+\frac{x^{[2\nu]}}{(2\nu)!}\Delta^{2\nu}f(-\nu)$$

$$=\frac{x^{[2\nu]-1}}{(2\nu-1)!}\,\frac{\Delta^{2\nu-1}f(-\nu)+\Delta^{2\nu-1}f(-\nu+1)}{2}+\frac{x^{[2\nu]}}{(2\nu)!}\Delta^{2\nu}f(-\nu).$$

It may, in a similar way, be derived from (15) alone. The three formulas (13), (15) and (17) produce, therefore, *if we stop at the order 2k, identical results*, but (17) has a theoretical advantage in the more symmetrical form, and in the fact that the function is split up into two parts, of which one is an *even*, the other an *odd* function of x.

29. The terms in brackets in (16) may be written

$$\frac{(x-\nu)\,x^{[2\nu]-1}}{(2\nu+1)!}\{(2\nu+1)\,\Delta^{2\nu}f(-\nu)+(x+\nu)\,[\Delta^{2\nu}f(-\nu+1)-\Delta^{2\nu}f(-\nu)]\}$$

$$=\frac{(x-\nu)\,x^{[2\nu]-1}}{(2\nu+1)!}[(x+\nu)\,\Delta^{2\nu}f(-\nu+1)-(x-\nu-1)\,\Delta^{2\nu}f(-\nu)]$$

$$=\frac{x^{[2\nu+2]-1}}{(2\nu+1)!}\delta^{2\nu}f(1)-\frac{(x-\nu)\,(x-\nu-1)\,x^{[2\nu]-1}}{(2\nu+1)!}\delta^{2\nu}f(0);$$

but we have

$$(x-\nu)\,(x-\nu-1)\,x^{[2\nu]-1}=(x+\nu-1)^{(2\nu+1)}=-(1-x)^{[2\nu+2]-1},$$

so that (16) assumes the form

$$\left.\begin{aligned} f(x) &= \sum_{0}^{k-1}\left[\frac{(1-x)^{[2\nu+2]-1}}{(2\nu+1)!}\delta^{2\nu}f(0)+\frac{x^{[2\nu+2]-1}}{(2\nu+1)!}\delta^{2\nu}f(1)\right]+R \\ R &= \frac{(x-k)\,x^{[2k]-1}}{(2k)!}f^{(2k)}(\xi). \end{aligned}\right\}\quad(19)$$

We shall call this formula Everett's first formula.[1]

[1] It had, however, already been found by Laplace, see G. J. Lidstone: Notes on Everett's Interpolation Formula (Proceedings of the Edinburgh Mathematical Society, vol. XL, 1921–22).

30. From (17) we obtain

$$\frac{f(x)+f(-x)}{2} = \sum_{\nu=0}^{k} \frac{x^{[2\nu]}}{(2\nu)!}\,\delta^{2\nu} f(0) + \frac{x^{[2k+2]}}{(2k+2)!} f^{(2k+2)}(\xi), \quad (20)$$

the remainder-term being derived by the following considerations. The exact form of the remainder-term in (17) is identical with the exact form of the remainder-term in (13) and (15), these formulas having the same remainder-term, viz.

$$R(x) = x^{[2k+2]-1} f(x, 0, \pm 1, \ldots \pm k) \quad (21)$$

whence

$$\frac{R(x)+R(-x)}{2} = x^{[2k+2]} \frac{f(x, 0, \pm 1, \ldots \pm k) - f(-x, 0, \pm 1, \ldots \pm k)}{2x}$$

$$= x^{[2k+2]} f(\pm x, 0, \pm 1, \ldots \pm k)$$

$$= \frac{x^{[2k+2]}}{(2k+2)!} f^{(2k+2)}(\xi)$$

or the remainder-term in (20).

31. Another class of Gaussian formulas is obtained as follows.

If, in (1), we put $a_{2\nu} = \nu + \frac{1}{2}$, $a_{2\nu+1} = -\nu - \frac{1}{2}$ and replace $f(t)$ by $f(t + \frac{1}{2})$, we obtain, stopping at the order $2k$,

$$\left.\begin{aligned} &f(x+\tfrac{1}{2}) = \\ &\sum_{0}^{k}\left[\frac{(x-\nu+\frac{1}{2})x^{[2\nu-1]-1}}{(2\nu-1)!}\Delta^{2\nu-1}f(-\nu+1)+\frac{x^{[2\nu+1]-1}}{(2\nu)!}\Delta^{2\nu}f(-\nu+1)\right]+R \\ &R = \frac{(x-k-\frac{1}{2})\,x^{[2k+1]-1}}{(2k+1)!} f^{(2k+1)}(\xi) \end{aligned}\right\} (22)$$

and, stopping at the order $2k - 1$,

$$\left.\begin{aligned} &f(x+\tfrac{1}{2}) = \\ &\sum_{0}^{k-1}\left[\frac{x^{[2\nu+1]-1}}{(2\nu)!}\Delta^{2\nu}f(-\nu+1)+\frac{(x-\nu-\frac{1}{2})\,x^{[2\nu+1]-1}}{(2\nu+1)!}\Delta^{2\nu+1}f(-\nu)\right]+R \\ &R = \frac{x^{[2k+1]-1}}{(2k)!} f^{(2k)}(\xi). \end{aligned}\right\} (23)$$

Next, putting $a_{2\nu} = -\nu - \frac{1}{2}$, $a_{2\nu+1} = \nu + \frac{1}{2}$ and stopping at the order $2k$, we find

$$\left.\begin{aligned} & f(x+\tfrac{1}{2}) = \\ & \sum_0^k \left[\frac{(x+\nu-\frac{1}{2})\,x^{[2\nu-1]-1}}{(2\nu-1)\,!}\Delta^{2\nu-1}f(-\nu+1) + \frac{x^{[2\nu+1]-1}}{(2\nu)\,!}\Delta^{2\nu}f(-\nu)\right] + R \\ & R = \frac{(x+k+\frac{1}{2})\,x^{[2k+1]-1}}{(2k+1)\,!}f^{(2k+1)}(\xi), \end{aligned}\right\} \tag{24}$$

while, stopping at the order $2k - 1$, we have

$$\left.\begin{aligned} & f(x+\tfrac{1}{2}) = \\ & \sum_0^{k-1} \left[\frac{x^{[2\nu+1]-1}}{(2\nu)\,!}\Delta^{2\nu}f(-\nu) + \frac{(x+\nu+\frac{1}{2})\,x^{[2\nu+1]-1}}{(2\nu+1)\,!}\Delta^{2\nu+1}f(-\nu)\right] + R \\ & R = \frac{x^{[2k+1]-1}}{(2k)\,!}f^{(2k)}(\xi). \end{aligned}\right\} \tag{25}$$

32. Taking, now, the arithmetical mean of (23) and (25), we find, by §2 (12),

$$\left.\begin{aligned} & f(x+\tfrac{1}{2}) = \sum_0^{k-1} \left[\frac{x^{[2\nu+1]-1}}{(2\nu)\,!}\,\square\,\delta^{2\nu}f(\tfrac{1}{2}) + \frac{x^{[2\nu+1]}}{(2\nu+1)\,!}\delta^{2\nu+1}f(\tfrac{1}{2})\right] + R \\ & R = \frac{x^{[2k+1]-1}}{(2k)\,!}f^{(2k)}(\xi). \end{aligned}\right\} \tag{26}$$

This formula which has also been found by Newton goes by the name of Bessel's formula. If we stop at the order $2k - 2$, we may, as remainder-term, use

$$R = \frac{x^{[2k-1]}}{(2k-1)\,!}f^{(2k-1)}(\xi_1) + \frac{x^{[2k+1]-1}}{(2k)\,!}f^{(2k)}(\xi_2). \tag{27}$$

In (26), use has been made of differences placed on the horizontal line starting from $f(\frac{1}{2})$, that is, mid-way between $f(0)$ and $f(1)$, together with arithmetical means of differences placed immediately above and below that line.

By a similar method as that which was employed for deriving (17) from (13) or from (15) alone, it may be shown that (26) can be derived from (23) or (25) alone, and that the three last-mentioned formulas, *if we stop at the order* $2k - 1$, *produce identical results.* On the other hand, Bessel's formula possesses similar theoretical advantages as Stirling's.

33. The difference in accuracy obtained by Stirling's and Bessel's formulas is, on the whole, but small. It may, therefore, be recommended as a practical rule to employ the former or latter of these formulas, according as we stop at differences of an even or odd order, thus avoiding remainder-terms of the forms (18) and (27).

34. The terms in brackets in (22) may, for $\nu > 0$, be written

$$\frac{(x-\nu+\frac{1}{2})\,x^{[2\nu-1]-1}}{(2\nu)\,!}\left[2\nu\,\Delta^{2\nu-1}f(-\nu+1)+(x+\nu-\tfrac{1}{2})\Delta^{2\nu}f(-\nu+1)\right]$$

$$=\frac{(x-\nu+\frac{1}{2})\,x^{[2\nu-1]-1}}{(2\nu)\,!}\left[(x+\nu-\tfrac{1}{2})\,\Delta^{2\nu-1}f(-\nu+2)-(x-\nu-\tfrac{1}{2})\,\Delta^{2\nu-1}f(-\nu+1)\right]$$

$$=\frac{x^{[2\nu+1]-1}}{(2\nu)\,!}\,\delta^{2\nu-1}f(\tfrac{3}{2})-\frac{(x-\nu+\frac{1}{2})\,(x-\nu-\frac{1}{2})\,x^{[2\nu-1]-1}}{(2\nu)\,!}\,\delta^{2\nu-1}f(\tfrac{1}{2});$$

but we have, for $\nu > 0$,

$$(x-\nu+\tfrac{1}{2})\,(x-\nu-\tfrac{1}{2})\,x^{[2\nu-1]-1}=(x+\nu-\tfrac{3}{2})^{(2\nu)}=(1-x)^{[2\nu+1]-1},$$

so that (22) may be written

$$\left.\begin{gathered}f(x+\tfrac{1}{2})=\\ f(1)+\sum_{1}^{k}\left[\frac{x^{[2\nu+1]-1}}{(2\nu)\,!}\,\delta^{2\nu-1}f(\tfrac{3}{2})-\frac{(1-x)^{[2\nu+1]-1}}{(2\nu)\,!}\,\delta^{2\nu-1}f(\tfrac{1}{2})\right]+R\\ R=\frac{(x-k-\frac{1}{2})\,x^{[2k+1]-1}}{(2k+1)\,!}\,f^{(2k+1)}(\xi).\end{gathered}\right\}\quad(28)$$

We shall refer to this formula as Everett's second formula.

35. From (26) we obtain

$$\frac{f(x+\frac{1}{2})-f(-x+\frac{1}{2})}{2}=\sum_{0}^{k-1}\frac{x^{[2\nu+1]}}{(2\nu+1)\,!}\delta^{2\nu+1}f(\tfrac{1}{2})+\frac{x^{[2k+1]}}{(2k+1)\,!}f^{(2k+1)}(\xi),\quad(29)$$

the remainder-term having been derived by the following considerations. The exact form of the remainder-term in (26) is identical with the exact form of the remainder-term in one of the formulas (23) or (25) which have the same remainder-term, viz.

$$R(x) = x^{[2k+1]-1} f(x + \tfrac{1}{2}, 0, \pm 1, \ldots \pm (k-1), k), \quad (30)$$

so that

$$\left.\begin{aligned} &\frac{R(x) - R(-x)}{2} = \\ &x^{[2k+1]} \frac{f(x+\frac{1}{2}, 0, \pm 1, \ldots \pm (k-1), k) - f(-x+\frac{1}{2}, 0, \pm 1, \ldots \pm (k-1), k)}{2x} \\ &= x^{[2k+1]} f(\pm x + \tfrac{1}{2}, 0, \pm 1, \ldots \pm (k-1), k) \\ &= \frac{x^{[2k+1]}}{(2k+1)!} f^{(2k+1)}(\xi), \end{aligned}\right\} (31)$$

being the remainder-term in (29).

If, in (29), we replace $f(t)$ by $f(t - \frac{1}{2})$, we obtain finally

$$\frac{f(x) - f(-x)}{2} = \sum_{0}^{k-1} \frac{x^{[2\nu+1]}}{(2\nu+1)!} \delta^{2\nu+1} f(0) + \frac{x^{[2k+1]}}{(2k+1)!} f^{(2k+1)}(\xi). \quad (32)$$

36. If we add (32) and (20) together, we get

$$\left.\begin{aligned} f(x) &= \sum_{0}^{n} \frac{x^{[\nu]}}{\nu!} \delta^{\nu} f(0) + R \\ R &= \frac{x^{[n+1]}}{(n+1)!} f^{(n+1)}(\xi_1) + \frac{x^{[n+2]}}{(n+2)!} f^{(n+2)}(\xi_2). \end{aligned}\right\} (33)$$

This is the *interpolation-formula with central differences;* it is of some theoretical importance, but not practically useful for interpolation purposes, as the differences of odd order are not immediately found in the difference-table.

37. A special case of particular interest is found by putting, in (26), $x = 0$. The result is the *formula for interpolation to halves,* or

$$\left.\begin{aligned} f(\tfrac{1}{2}) &= \sum_{0}^{k-1} (-1)^{\nu} \frac{[1 \cdot 3 \ldots (2\nu - 1)]^2}{2^{2\nu}(2\nu)!} \square\, \delta^{2\nu} f(\tfrac{1}{2}) + R \\ R &= (-1)^{k} \frac{[1 \cdot 3 \ldots (2k-1)]^2}{2^{2k}(2k)!} f^{(2k)}(\xi). \end{aligned}\right\} (34)$$

38. On many occasions not only the function but also its differential coefficient is known at a number of given points. If both of these values are used in the interpolation, it is called *osculating interpolation* (of the first order). Osculation of higher order occurs, if differential coefficients of higher order are introduced. The problem has, as a matter of fact, already been solved by (1) which is also valid, if some of the points a_ν coincide. We simply have to interpret the divided differences with repeated arguments, thus introduced, in accordance with the principles of No. 20, 7°.

If, for instance, the values of $f(a)$, $f'(a)$, $f''(a)$ and $f(b)$ are given, we also know $f(a, a) = f'(a)$ and $f(a, a, a) = \frac{1}{2} f''(a)$, so that we can form the difference-table in No. 20, 7°, whereafter we have

$$f(x) = f(a) + (x-a) f(a,a) + (x-a)^2 f(a,a,a) + (x-a)^3 f(a,a,a,b) + R,$$

$$R = (x - a)^3 (x - b) \frac{f^{(4)}(\xi)}{24}.$$

39. If, on the other hand, we use Lagrange's instead of Newton's formula, the matter is not quite so simple, as the separate terms in §3(16) may tend to infinity, if some of the points a_ν coincide. In that case, the terms tending to infinity together must be collected into one term and the limiting value be examined, while the points that are finally to coincide must be kept separate during the limiting process. If, for instance, a_ν, a_μ and a_λ are ultimately to coincide, we may put $a_\nu = a$, $a_\mu = a + \epsilon$, $a_\lambda = a + 2\epsilon$ and let ϵ tend to zero.

We do not propose to do this in detail, but content ourselves with stating an important result and verifying it. Putting

$$f(x) = \sum_0^n \frac{P_\nu^2(x)}{P_\nu^2(a_\nu)} f(a_\nu) + \sum_0^n \frac{P_\nu(x)\, P(x)}{P_\nu^2(a_\nu)} \left[f'(a_\nu) - 2 \frac{P_\nu'(a_\nu)}{P_\nu(a_\nu)} f(a_\nu) \right], \quad (35)$$

$P(x)$ and $P_\nu(x)$ having the same meaning as in §3 (15), we intend to show that this expression represents the polynomial of degree $2n + 1$ which for $x = a_0, a_1, \ldots a_n$ assumes the values $f(a_0)$, $f(a_1), \ldots f(a_n)$, while $f'(x)$ assumes the values $f'(a_0)$, $f'(a_1), \ldots f'(a_n)$.

It is, to begin with, clear that for $x = a_r$, r having one of the values $0, 1, \ldots n$, (35) assumes the value $f(a_r)$; for each separate term in the second sum vanishes on account of the factor $P(x)$, while of the first sum only the term $f(a_r)$ with the factor 1 remains.

Differentiating now (35) and putting, thereafter, $x = a_r$, we get

$$f'(a_r) = \sum_0^n \frac{2P_\nu(a_r)\,P'_\nu(a_r)}{P_\nu^2(a_\nu)} f(a_\nu) + \sum_0^n \frac{P_\nu(a_r)\,P'(a_r)}{P_\nu^2(a_\nu)}\left[f'(a_\nu) - 2\frac{P'_\nu(a_\nu)}{P_\nu(a_\nu)} f(a_\nu)\right]$$

or, as $P_\nu(a_r) = 0$ for $\nu \neq r$,

$$f'(a_r) = \frac{2\,P'_r(a_r)}{P_r(a_r)} f(a_r) + \frac{P'(a_r)}{P_r(a_r)}\left[f'(a_r) - 2\frac{P'_r(a_r)}{P_r(a_r)} f(a_r)\right];$$

but the correctness of this equation follows from the fact, that $P'(a_r) = P_r(a_r)$, as is seen by differentiating $P(x) = (x - a_r)\,P_r(x)$ and putting $x = a_r$.

If (35) is applied to a function which is not a polynomial, it becomes necessary to add the remainder-term which, according to what has been said in No. 24, is identical with the remainder-term in Newton's formula under the same conditions, that is

$$R = P^2(x)\,f(x, a_0, a_0, \ldots a_n, a_n) = P^2(x)\,\frac{f^{(2n+2)}(\xi)}{(2n+2)\,!}. \qquad (36)$$

§5. Some Applications

40. Before proceeding to show, by means of numerical examples, how the interpolation-formulas developed in §4 are applied in practice, we find it advisable to illustrate the warning, given in No. 2, against indiscriminate application of polynomials for interpolation-purposes. We choose a case where intuition proves of no value as a protection against erroneous conclusions.

Let us assume, that we have to do with a function which, in a given interval $-a \leq x \leq a$, does not present any "peculiarities" whatever; we may, in order to put it more precisely, assume that, in this interval, the function is everywhere continuous and possesses differential coefficients of every order. Most practical computers will, then, be inclined to believe that any desired degree of agreement between the polynomial used for the interpolation, and the function, may, within the interval, be obtained by arranging that the polynomial assumes the same values as the function at a sufficient number of points which are uniformly distributed over the interval.

Yet this conclusion is unwarranted, as we proceed to show by an example.

We consider the function

$$f(x) = \frac{1}{1 + x^2}, \tag{1}$$

a function which is evidently continuous and possesses differential coefficients of every order in every finite interval. This function is so simple that a theory of interpolation with any claim to authority should be able to handle it safely.

In order to form the divided differences it is convenient to write the function in the form

$$f(x) = \frac{i}{2}\left(\frac{1}{i - x} + \frac{1}{i + x}\right), \tag{2}$$

where $i = \sqrt{-1}$. As evidently also for complex z

$$\theta_r\theta_{r-1} \cdots \theta_1 \frac{1}{z - x_0} = \frac{1}{(z - x_0)(z - x_1) \cdots (z - x_r)}, \tag{3}$$

we have

$$f_r = \frac{i}{2}\left[\frac{1}{(i - x_0) \cdots (i - x_r)} + \frac{(-1)^r}{(i + x_0) \cdots (i + x_r)}\right]. \tag{4}$$

We now assume that the values of $f(x)$ have been given at a number of points which are uniformly distributed over the interval $-5 \leq x \leq 5$, and propose to examine the result of interpolating by Newton's formula to $f(x)$, x being a point within the said interval. As our only object is to show that there are cases where this interpolation fails, and that the failure is the greater the more values of the function are used, we may choose the uniformly distributed points in a special way with a view to obtaining the simplest possible calculations. We assume, therefore, that the values of the function are given at the points

$$\pm\frac{1}{\nu}, \pm\frac{3}{\nu}, \pm\frac{5}{\nu}, \cdots \pm 5 \qquad (5\nu + 1 \text{ points}), \tag{5}$$

so that the interval has been divided into 5ν equal parts, *ν being an odd number of the form $4k - 1$.* We have, therefore, $\nu \geq 3$.

The problem is to examine what happens to the remainder-term in Newton's formula if, for ν, we insert ever larger values.

It is easy to calculate this remainder-term. For we have

$$R = P(x)f\left(x, \pm\frac{1}{\nu}, \pm\frac{3}{\nu}, \cdots \pm 5\right) \tag{6}$$

where

$$\left.\begin{aligned} P(x) &= \left(x^2 - \frac{1}{\nu^2}\right)\left(x^2 - \frac{9}{\nu^2}\right)\left(x^2 - \frac{25}{\nu^2}\right)\cdots(x^2 - 25) \\ &= (x + 5)\left(x + 5 - \frac{2}{\nu}\right)\left(x + 5 - \frac{4}{\nu}\right)\cdots(x - 5) \end{aligned}\right\} \tag{7}$$

and, by (4),

$$\begin{aligned} f\left(x, \pm\frac{1}{\nu}, \pm\frac{3}{\nu}, \cdots \pm 5\right) &= \frac{i}{2}\left[\frac{1}{(i - x)\,P(i)} + \frac{(-1)^{5\nu+1}}{(i + x)\,P(i)}\right] \\ &= \frac{1}{(1 + x^2)\,P(i)}, \end{aligned}$$

so that

$$R = \frac{1}{1 + x^2}\cdot\frac{P(x)}{P(i)}, \tag{8}$$

but, as $\nu = 4k - 1$, $\dfrac{5\nu + 1}{2}$ must be an even number; therefore

$$P(i) = \left(1 + \frac{1}{\nu^2}\right)\left(1 + \frac{9}{\nu^2}\right)\left(1 + \frac{25}{\nu^2}\right)\cdots(1 + 25). \tag{9}$$

If, for abbreviation, we put

$$x = 2z - 5,\ z = \frac{x + 5}{2}, \tag{10}$$

we obtain, by (7),

$$\begin{aligned} P(x) = P(2z - 5) &= \left(\frac{2}{\nu}\right)^{5\nu+1} \nu z(\nu z - 1)\cdots(\nu z - 5\nu) \\ &= \left(\frac{2}{\nu}\right)^{5\nu+1} \frac{\Gamma(1 + \nu z)}{\Gamma(\nu z - 5\nu)} \end{aligned}$$

or, as $\Gamma(t)\Gamma(1 - t) = \dfrac{\pi}{\sin \pi t}$,

$$P(x) = -\left(\frac{2}{\nu}\right)^{5\nu+1} \frac{\sin \nu\pi z}{\pi}\,\Gamma(1 + \nu z)\Gamma(1 + 5\nu - \nu z).$$

As $0<z<5$, owing to $-5< x <5$, we may apply the well-known inequality, valid for $t>0$,

$$\Gamma(1+t) > t^t e^{-t}\sqrt{2\pi t}$$

and thus obtain

$$|P(x)| > 4\,|\sin \nu z\pi|\,\sqrt{z(5-z)}\left[\left(\frac{2}{e}\right)^5 z^z(5-z)^{5-z}\right]^\nu. \tag{11}$$

Remembering, now, that the geometrical mean is smaller than the arithmetical, we have

$$[P(i)]^{\frac{2}{5\nu+1}} < \frac{\left(1+\frac{1}{\nu^2}\right)+\left(1+\frac{3^2}{\nu^2}\right)+\cdots+\left(1+\frac{(5\nu)^2}{\nu^2}\right)}{\frac{5\nu+1}{2}}$$

$$= 1+\frac{2}{5\nu+1}\cdot\frac{1}{\nu^2}(1+3^2+5^2+\cdots+(5\nu)^2);$$

but

$$1+3^2+5^2+\cdots+(2k+1)^2 = \tfrac{1}{6}(2k+1)(2k+2)(2k+3),$$

as is proved by induction or otherwise; therefore

$$[P(i)]^{\frac{2}{5\nu+1}} < 1+\frac{2}{5\nu+1}\cdot\frac{1}{\nu^2}\cdot\frac{1}{6}\,5\nu(5\nu+1)(5\nu+2)$$

$$= \frac{28}{3}+\frac{10}{3\nu}$$

or

$$P(i) < \left(\frac{28}{3}+\frac{10}{3\nu}\right)^{\frac{5\nu+1}{2}}$$

$$= \left(\frac{28}{3}\right)^{\frac{5\nu+1}{2}}\left(1+\frac{5}{14\nu}\right)^{\frac{5\nu}{2}}\sqrt{1+\frac{5}{14\nu}}.$$

The last factor we replace by $\sqrt{1+\frac{5}{42}}=\sqrt{\frac{47}{42}}$, being at least as large, since $\nu\geq 3$. Further, we have for $t>0$

$$1+\frac{t}{\nu} < e^{\frac{t}{\nu}},$$

as is seen by developing in powers of t, so that the middle factor may be replaced by $e^{\frac{25}{28}}$. We thus have

$$P(i) < \frac{\sqrt{94}}{3}\left(\frac{28}{3}\right)^{\frac{5\nu}{2}} e^{\frac{25}{28}}. \tag{12}$$

But from (12) and (11) jointly with (8), it follows, that

$$|R| > \frac{1}{1+x^2}\,\frac{12\,|\sin \nu z\pi|\,\sqrt{z(5-z)}}{\sqrt{94}\,e^{\frac{25}{28}}}\left[\frac{\left(\frac{2}{e}\right)^5 z^z(5-z)^{5-z}}{\left(\frac{28}{3}\right)^{\frac{5}{2}}}\right]^\nu. \tag{13}$$

We are now in a position to examine what happens to R for special values of x or z inside the interval considered. As an example, we will put $x = 4\frac{7}{8}$, that is $z = 4\frac{15}{16}$. We have for this particular value, as $\nu = 4k - 1$,

$$|\sin \nu z\pi| = \left|\sin(4k-1)\left(5 - \frac{1}{16}\right)\pi\right|$$

$$= \left|\sin\left(\frac{k}{4} - \frac{1}{16}\right)\pi\right|;$$

but this expression cannot, k being an integer, assume any other values than $\sin\frac{\pi}{16}$, $\sin\frac{3\pi}{16}$, $\sin\frac{5\pi}{16}$, $\sin\frac{7\pi}{16}$, amongst which we select the smallest, being $\sin\frac{\pi}{16}$. A numerical calculation shows, finally, that for $x = 4\frac{7}{8}$

$$|R| > \frac{(1.8)^\nu}{451},$$

so that, with increasing ν, $|R|$ increases beyond any limit.

The determination of the interval within which the process is *convergent* is due to Runge,[1] but requires analytical methods which are beyond the scope of this book.

This result, of course, does not mean that it is not permitted to interpolate in a table of the function $f(x) = \frac{1}{1+x^2}$, but only that the degree of accuracy obtained should not be estimated by intuition, but by means of the remainder-term.

[1] This interval is $\pm 3{\cdot}63\ldots$; see, for instance, P. Montel: Leçons sur les séries de polynomes, p. 51–55.

We learn by this example that even extremely simple cases exist where the agreement between the function and the polynomial, used for the interpolation, at a number of given points 0, 1, 2, . . . n, is obtained only at the cost of very large deviations between function and polynomial in the intervals between these points. We refer, as an illustration, to figure 1 where the dotted line indicates a polynomial of this nature.*

41. We now approach the numerical applications of the interpolation-formulas. If tables of the coefficients in these formulas[1] with a sufficient number of decimals and proceeding by sufficiently

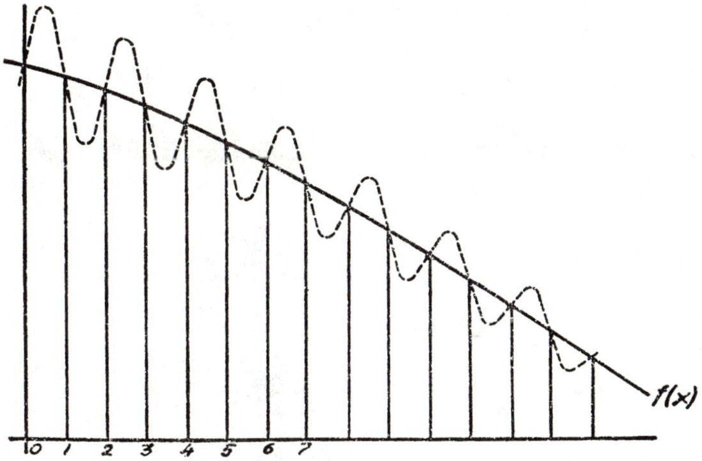

FIG. 1

small intervals, are available, and particularly if an arithmometer is at hand, the formulas may be left as they stand in §4. The only point requiring further comment, is the practical use of the remainder-term. In other cases it may be recommended to write

[1] See, for instance, J. W. Glover: Tables of Applied Mathematics, Ann Arbor, Michigan, 1923, p. 412; A. J. Thompson: Table of the Coefficients of Everett's Central-Difference Interpolation Formula (Tracts for Computers, No. V).

the formulas in a slightly different way. For instance, Newton's formula with divided differences should be written

$$f(x) = f_0 + (x - a_0)\,[f_1 + (x - a_1)\,[f_2 + (x - a_2)\,[f_3 + \ldots . \quad (14)$$

where, as in the following formulas, the remainder-term has been left out. The idea is, that the calculation should be commenced from the right with the difference of the highest order, say, f_3; this is multiplied by $(x - a_2)$; to the product is added f_2; the result is multiplied by $(x - a_1)$; and so on.

The other formulas are best written in symbolical form, leaving out the symbolical factor $f(0)$ or $f(\frac{1}{2})$ on both sides. Thus, the formula with descending differences may be written

$$E^x = 1 + x\left(\Delta + \frac{x-1}{2}\left(\Delta^2 + \frac{x-2}{3}\left(\Delta^3 + \frac{x-3}{4}\left(\Delta^4 + \ldots, \quad (1\right.\right.\right.\right.$$

the factor $f(0)$ having been left out. The formula with ascending differences becomes

$$E^x = 1 + x\left(\nabla + \frac{x+1}{2}\left(\nabla^2 + \frac{x+2}{3}\left(\nabla^3 + \frac{x+3}{4}\left(\nabla^4 + \ldots, \quad (1\right.\right.\right.\right.$$

and Stirling's formula

$$\left.\begin{aligned} E^x &= 1 + \frac{x^2}{1.2}\left(\delta^2 + \frac{x^2-1}{3.4}\left(\delta^4 + \frac{x^2-4}{5.6}\left(\delta^6 + \ldots\right)\right)\right) \\ &\quad + x\left(\square\,\delta + \frac{x^2-1}{2.3}\left(\square\,\delta^3 + \frac{x^2-4}{4.5}\left(\square\,\delta^5 + \ldots;\right.\right.\right. \end{aligned}\right\}(17)$$

in all of these the factor left out is $f(0)$.

On the other hand, in the case of Bessel's formula

$$\left.\begin{aligned} E^x &= \square + \frac{x^2-\frac{1}{4}}{1.2}\left(\square\,\delta^2 + \frac{x^2-\frac{9}{4}}{3.4}\left(\square\,\delta^4 + \frac{x^2-\frac{25}{4}}{5.6}\left(\square\,\delta^6 + \ldots\right)\right)\right) \\ &\quad + x\left(\delta + \frac{x^2-\frac{1}{4}}{2.3}\left(\delta^3 + \frac{x^2-\frac{9}{4}}{4.5}\left(\delta^5 + \ldots\right.\right.\right. \end{aligned}\right\}(1$$

the factor left out is $f(\frac{1}{2})$.

Everett's first formula may, if $x + y = 1$, be written

$$\left.\begin{aligned} E^x &= y\left(1 + \frac{y^2-1}{2.3}\left(\delta^2 + \frac{y^2-4}{4.5}\left(\delta^4 + \dots\right)\right)\right) \\ &+ x\left(E + \frac{x^2-1}{2.3}\left(\delta^2 E + \frac{x^2-4}{4.5}\left(\delta^4 E + \dots\right.\right.\right. \end{aligned}\right\} \quad (19)$$

where the factor $f(0)$ has been left out; and Everett's second formula

$$\left.\begin{aligned} E^x &= E^{\frac{1}{2}} + \frac{x^2-\frac{1}{4}}{1.2}\left(\delta E + \frac{x^2-\frac{9}{4}}{3.4}\left(\delta^3 E + \dots\right)\right) \\ &\qquad - \frac{y^2-\frac{1}{4}}{1.2}\left(\delta + \frac{y^2-\frac{9}{4}}{3.4}\left(\delta^3 + \dots\right.\right. \end{aligned}\right\} \quad (20)$$

the factor $f(\frac{1}{2})$ having been left out.

42. Everett's first formula is particularly useful in connection with tables of functions which, according to a practice introduced by K. Pearson,[1] have been tabulated for rather wide intervals but are accompanied by auxiliary tables of δ^2 and δ^4. In such tables, space is saved by leaving out δ and δ^3, and interpolation by Everett's formulas is, as regards the work implied, practically equivalent to interpolation by Stirling's or Bessel's formulas.

As regards the other formulas, interpolation with descending or ascending differences is necessary at the beginning and end of a table, as central differences are not to be had there. Linear interpolation would, of course, never be done by Bessel's formula, but by the descending or ascending first difference. Also in the case of interpolation with second differences there is, as a rule, little advantage in using central differences. Otherwise Stirling's or Bessel's formula is usually applied. If the intervals are not equal, or if some of the central differences, but not of sufficiently high order, are obtainable, interpolation with divided differences may be resorted to. In that case, in order not to compute with higher numbers than necessary, the arguments $a_0, a_1, a_2, \ldots$ in (14) are chosen in such an order that a_0 is nearest to x, a_1 after that nearest to

[1] Tracts for Computers, Nos. II and III: On the Construction of Tables and on Interpolation.

x, etc. The divided differences used in the calculation will then be situated on a zig-zag-line, starting as close as possible to x.

43. We will first deal with a case where the given values of $f(x)$ are *exact*, not approximate numbers.

x	$f(x)$	Δ	Δ^2	Δ^3
2.1	427.8582			
		110.9306		
2.2	538.7888		21.8990	
		132.8296		3.0930
2.3	671.6184		24.9920	
		157.8216		
2.4	829.4400			

The table above is part of a table of the function $x^4 + 10x^5$. It is required, by interpolation to calculate $f(2.14)$ and, by means of the remainder-term, to indicate limits for the error committed.

We commence by forming the difference-table as shown, choose 2.1 as origin and note that the unit of interval is 0.1, so that, in (15), we must put $x = 0.4$. The details of the calculation are then as follows:

$$\begin{array}{rl}
\dfrac{0.4-2}{3}\Delta^3 = - & 1.6496 \\
+\Delta^2 & 21.8990 \\
& 20.2494 \times \dfrac{0.4-1}{2} \\
= & -\,6.07482 \\
+\Delta & 110.9306 \\
& 104.85578 \times 0.4 \\
= & 41.942312 \\
+f(2.1) & 427.8582 \\
& 469.800512 = f(2.14).
\end{array}$$

In order to calculate the remainder-term we put

$$f(x) = f(2.1 + 0.1\,x') = F(x') = \frac{(x' + 21)^4 + (x' + 21)^5}{10000},$$

and find for the remainder-term, ξ being comprised between 0 and 3,

$$R = \frac{0.4^{(4)}}{4\,!} F^{(4)}(\xi) = \frac{0.4^{(4)}}{4\,!} \cdot \frac{4\,! + 5^{(4)}(\xi + 21)}{10000}$$

$$= -\,0.01058304 - 0.0004992\,\xi.$$

If this is added to the value found for $f(2.14)$, we finally find

$$f(2.14) = 409.78992896 - 0.0004992\,\xi \quad (0 < \xi < 3).$$

It is seen that $f(2.14)$ is comprised between 469.790 and 469.788. The correct value to 4 decimals is, in fact, 469.7893.

It should be noted that in this case the remainder-term contains a constant part which to a considerable extent contributes towards improving the value immediately found by the interpolation.

44. In practice, it is advisable to begin by examining the remainder-term, in order to ascertain how many decimals it is necessary to take into account in the calculation. Thus, an examination of the above expression for R, shows that the uncertainty caused by ξ can only amount to $1\frac{1}{2}$ units of the 3rd decimal, so that, in practice, only four decimals should be retained in the course of the calculation. We have, however, in this and in the following example, preferred to carry out the calculation with the exact values of the function, in order to avoid, for the moment, other sources of error than the remainder-term itself.

45. If, in the same table, we want to interpolate for $f(2{\cdot}27)$, central differences should be used, and as we know a δ^3, but no $\Box\,\delta^3$, it is Bessel's formula that should be used, in agreement with the rule suggested in No. 33: to apply Bessel's formula, if we stop at a difference of odd order.[1] The calculation is performed in the following way. We first calculate for the middle interval $\Box\,f(2.25) = 605.2036$ and $\Box\,\delta^2 f(2.25) = 23.4455$; with a little practice this can be done mentally, and the results be inserted in their places in the difference-table, whereafter all the figures to be used in the calculation are placed on the same horizontal line. There-

[1] We might, of course, also apply one of the Gaussian formulas, in a form analogous to (15).

after, we put in the formula $x = 0.2$, being the distance from the middle of the interval, and find in succession

$$\frac{0.04-0.25}{2}\Box\,\delta^2 = -\ 2.4617775 \qquad \frac{0.04-0.25}{6}\,\delta^3 = -\ 0.108255$$

$$\begin{array}{llll} +\ \Box & 605.2036 & +\ \delta & 132.8296 \\ & 602.7418225 & & 132.721345 \times 0.2 \\ & & = & 26.544269 \\ & & + & 602.7418225 \\ & & & 629.2860915 = f(2.27) \end{array}$$

For the remainder-term we find, ξ being comprised between $\pm \frac{3}{2}$,

$$R = \frac{(0.2)^{[5]-1}}{4\,!}\cdot\frac{4\,! + 5^{(4)}\,(\xi + 22.5)}{10000} = 0.005267535 + 0.00023205\,\xi,$$

so that $f(2.27)$ is comprised between 629.292 and 629.291. The correct value to 4 places is 629.2914.

46. If, for the same interpolation, we apply one of Everett's formulas, it is clear that it is Everett's first formula that must be applied, as we know two differences of the second order but only one of the third. As both differences of the second order are used, it appears that all the information contained in the difference-table is turned to account, although we apparently stop at the second order of differences. The details of the calculation are in practice as follows. We choose 2.2 as origin and put $x = 0.7$, $y = 0.3$. The remainder-term is

$$R = \frac{0.7\,(0.49 - 1)\,(0.7 - 2)}{4\,!}\cdot\frac{4\,! + 5^{(4)}\,(\xi + 22)}{10000}$$

$$= 0.00515151 + 0.00023205\,\xi \quad (-\ 1 < \xi < 2).$$

The uncertainty caused by ξ can, therefore, amount to 7 units of the fourth place of decimals. The third decimal being thus uncertain, we need only retain four decimals in the calculation. The successive steps in this are:

$\frac{0.09-1}{6}\,\delta^2 =$	$-\ 3.3213$	$\frac{0.49-1}{6}\,\delta^2 E =$	$-\ 2.1243$
$+f(2.2)$	538.7888	$+f(2.3)$	671.6184
	535.4675×0.3		669.4941×0.7
	$= 160.6402$		$= 468.6459$
			$+\ 160.6402$
			$629.2861 = f(2.27)$

If to this we add the remainder-term

$$R = 0.0052 + 0.00023\,\xi,$$

we have as the result of the interpolation

$$f(2.27) = 629.2913 + 0.00023\,\xi \quad (-1 < \xi < 2),$$

so that the required value is found, as before, to be comprised between 629.292 and 629.291.

It is seen that the actual work of computing is about the same as in the case of Bessel's formula, taking into account the forming of arithmetical means in the latter case.

47. Before proceeding to the case where the given values of $f(x)$ are not exact but approximate numbers, it may be noted that whether they are exact or approximate, it is practical to begin with testing the table for grosser errors, caused by mistakes in writing, printing or calculating, etc. This is done by mere inspection of the difference-table which must be formed anyhow.

Let us consider the effect on the difference-table of a single error of one unit. Its propagation can be studied by means of the following difference-table of the errors which is, at the same time, a table of the errors in the differences of the function.

Error	Δ	Δ^2	Δ^3	Δ^4	Δ^5	Δ^6
0		0		0		1
	0		0		1	
0		0		1		−6
	0		1		−5	
0		1		−4		15
	1		−3		10	
1		−2		6		−20
	−1		3		−10	
0		1		−4		15
	0		−1		5	
0		0		1		−6
	0		0		−1	
0		0		0		1

The numbers in this table are binomial numbers, as appears from the expression of $\Delta^n f(x)$ if, in this, we put $f(0) = 1, f(\nu) = 0$ $(\nu \neq 0)$. If, now, we form the difference-table of $f(x)$ and carry it sufficiently far for the differences, considering the number of places to which we work, to vanish, the effect of an isolated error will be that instead of a column of zeros we get a column of binomial numbers, multiplied by the same constant k (if the error is k instead of 1). At the same time this column shows which value of $f(x)$ is wrong, and the approximate magnitude of the error.

Let us, for instance, consider the following section of a table of logarithms.

x	$\log x$	Δ	Δ^2
3.8500	0.5854607		
		113	
8501	4720		0
		113	
8502	4833		−6
		107	
8503	4940		11
		118	
8504	5058		−5
		113	
8505	5171		0
		113	
8506	5284		0
		113	
8507	5397		

Inspection of the column of Δ^2 shows at once that log 3.8503 is erroneous, and that the error is approximately −6 units of the last decimal (the correct value is 0.5854946).

48. We now assume that this kind of error is not present, and that the only kind of error in the functional values is due to their having been *rounded off* to the nearest unit of the last decimal retained. An error of this nature can, at most, amount to $\frac{1}{2}$ unit of the last figure, and an upper limit to the error in the differences is therefore obtained by considering the following table:

Error	Δ	Δ^2	Δ^3	Δ^4	Δ^5
−0.5					
	1				
0.5		−2			
	−1		4		
−0.5		2		−8	
	1		−4		16
0.5		−2		8	
	−1		4		
−0.5		2			
	1				
0.5					

It is seen that the numerical value of the error in Δ^n cannot exceed 2^{n-1} units of the kind considered. It follows, that we shall generally be guarded against the influence of this kind of error by carrying out the calculation with one or two more figures than we need in the result. As a rule the influence of these "forcing-errors," as we shall call them, is considerably smaller than would appear from the table above, as they counterbalance each other to a certain extent. They are of the same nature, and are dealt with in the same way, as the errors introduced at the various steps of the interpolation where approximations are resorted to; and the practical computer will all the time be conscious of the number of figures to which these approximations must be taken in order not to lose any of the accuracy with which he started. Apart from the cases where very long calculations are required in order to reach the result, the forcing-errors are not very dangerous, especially in comparison with the errors into which the computer may fall by underrating the influence of the remainder-term.

In most tables the distribution of the forcing-errors may be considered "accidental" in the sense in which this word is taken in the theory of probabilities. On this assumption, it is theoretically possible to determine the mean error[1] of the interpolated value of $f(x)$, a problem which, however, is not of great practical importance.

In practice, the forcing-errors in the given values of the function show their presence in the difference-table. For, if we consider a column of differences which, taking the nature of the function and the degree of accuracy into account, ought to vanish, it will be

[1] Thiele: Interpolationsrechnung, p. 38–42.

found that instead of vanishing it presents frequent changes of sign, while a not too large section of it will approximately have the sum zero. In that case, differences of this order are not used in the interpolation, and the remainder-term will often show that it is possible to stop at a still lower difference.

49. As an example we will consider a section of a table of $\sqrt{x}$.

x	$\sqrt{x}$	Δ	Δ^2	Δ^3	Δ^4	Δ^5
5451	73.8308879					
		0.3378387				
5501	74.1687266		−15319			
		0.3363068		207		
5551	74.5050334		−15112		−6	
		0.3347956		201		3
5601	74.8398290		−14911		−3	
		0.3333045		198		−1
5651	75.1731335		−14713		−4	
		0.3318332		194		−3
5701	75.5049667		−14519		−7	
		0.3303813		187		
5751	75.8353480		−14332			
		0.3289481				
5801	76.1642961					

The differences from Δ^2 on have been stated in units of the last decimal, as is often done to save space.

According to the nature of the function, the nth differential coefficient and, consequently, the nth difference should have the constant sign $(-1)^{n-1}$. The figures show, however, that the sign of Δ^5 changes, and the sum of this column is very small (one unit of the last decimal). There is, therefore, no reason to use Δ^5 in the interpolation; whether Δ^4 can also be left out depends on the remainder-term.

As an example, let us interpolate for $\sqrt{5616}$. If, in Everett's first formula, we put $x = 0.3$, we find for the remainder-term, stopping at second differences,

$$R = \frac{0.3\ (0.09 - 1)\ (0.3 - 2)}{4!} D^4\,(5601 + 50\xi)^{\frac{1}{2}}\ (-1 < \xi < 2),$$

and a short calculation shows that this can at most affect the result with one unit of the 8th decimal. We may, therefore, ignore the remainder-term, using the formula to second differences which is, in fact, equivalent to going to the third difference in the table, as the formula uses the second differences of both $f(0)$ and $f(1)$.

The calculation proceeds as follows:

$$\frac{0.49-1}{6}\delta^2 = 0.00012674 \qquad \frac{0.09-1}{6}\delta^2 E = 0.00022315$$

$$\begin{array}{llll} +\sqrt{5601} & 74.8398290 & +\sqrt{5651} & 75.1731335 \\ & 74.83995574 \times 0.7 & & 75.17335665 \times 0.3 \\ & = 52.38796902 & & = 22.55200700 \\ & & & + 52.38796902 \\ & & & 74.9399760 = \sqrt{5616} \end{array}$$

The result is correct to the last decimal. The setting out of all the differences in the table might, of course, have been avoided by beginning with the examination of the remainder-term.

We may use the same table for illustrating the use of the Error-Test. Confining ourselves to the case of linear interpolation, we use the arguments 0 and 1, and calculate $f(x)$ by

$$f(x) = f(0) + x \triangle f(0),$$

the remainder-term being $\frac{1}{2}x(x-1)f''(\xi)$. If, as the next argument, we take 2, the remainder-term is $\frac{1}{6}x(x-1)(x-2)f'''(\xi)$, but for $0 < x < 1$ the two remainder-terms have the same sign, as f'' and f''' have opposite signs for $f(x) = \sqrt{x}$, and the Error-Test is inapplicable. If, on the other hand, we take -1 as the next argument, the remainder-term is $\frac{1}{6}x(x^2-1)f'''(\xi)$ which for $0 < x < 1$ has the opposite sign of $\frac{1}{2}x(x-1)f''(\xi)$. With this choice of arguments the first neglected term in the linear interpolation-formula is $\frac{1}{2}x(x-1)\triangle^2 f(-1)$, and the error is, therefore, numerically less than this expression and has the same sign.

Performing the calculation for $x = 0.3$ we find

$$\sqrt{5616} = 74.8398290 + 0.3 \times 0.3333045$$

$$= 74.9398204$$

with an error that is numerically less than

$$\frac{0.3\,(-0.7)}{2}(-0.0014911) = 0.0001566$$

and has the same sign: that is, we have

$$74.9398204 < \sqrt{5616} < 74.9399770.$$

Of the two limits, the one obtained by adding the next term is, as might be expected, much the better. But if we want to act in absolute safety, the result of the interpolation should be stated as $\sqrt{5616} = 74.9399$ with a possible error of one unit of the fourth decimal.

50. As an application of Stirling's formula, let us consider the table

x	e^{-x}	Δ	Δ^2	Δ^3	Δ^4
0.1	0.90484				
		−8611			
0.2	0.81873		820		
		−7791		−79	
0.3	0.74082		741		9
		−7050		−70	
0.4	0.67032		671		
		−6379			
0.5	0.60653				

We propose to interpolate for $e^{-0.34}$, making use of the fourth difference. In the formula we must, then, put $x = 0.4$, and find, for the remainder-term,

$$R = \frac{-0.4\,(0.16 - 1)\,(0.16 - 4)}{5!\,10^5}\,e^{-\frac{3+\xi}{10}} \quad (-2 < \xi < 2);$$

it is seen, without actual calculation, that it cannot influence the result in which only 5 decimals are to be retained. The details of the calculation are:

$$\begin{array}{lll} \dfrac{0.16-1}{12}\,\delta^4 = & -0.000006 & \\ +\,\delta^2 & 0.00741 & \\ & 0.007404 \times \dfrac{0.16}{2} & \\ = & 0.000592 & \\ +\,e^{-0.3} & 0.74082 & \\ & 0.741412 & \end{array} \qquad \begin{array}{lll} \dfrac{0.16-1}{6}\,\square\,\delta^3 = & 0.000104 & \\ +\,\square\,\delta & -0.074205 & \\ & -0.074101 \times 0.4 & \\ = & -0.029640 & \\ + & 0.741412 & \\ & 0.71177 = e^{-0.34} & \end{array}$$

the result being correct to the last place.

It is seen that δ^4 has no influence, and it would, therefore, have been sufficient to use Bessel's formula to the third difference. What we have done, is practically equivalent to using Stirling's formula to the third difference and with a remainder-term of the double form §4 (18); as the trouble in ascertaining that $\frac{x^{[4]}}{4!} f^{(4)}(\xi)$ is also without influence, may be compared with the trouble we had in calculating the term involving δ^4.

51. If we treat the same case by Everett's second formula, we must put $x = 0.9$, $y = 0.1$ and find

$$
\begin{aligned}
\frac{0.81 - 2.25}{12}\delta^3 E &= \quad 0.000084 \\
+ \delta E \quad & \quad - 0.07050 \\
& \quad - 0.070416 \times \frac{0.81 - 0.25}{2} \\
&= - 0.019716 \\
+ e^{-0.3} \quad & \quad\ \ 0.74082 \\
& \quad\ \ 0.721104
\end{aligned}
$$

$$
\begin{aligned}
\frac{0.01 - 2.25}{12}\delta^3 &= \quad 0.000147 \\
+ \delta \quad & \quad - 0.07791 \\
& \quad - 0.077763 \times \frac{-0.01 + 0.25}{2} \\
&= - 0.009332 \\
&+ \quad 0.721104 \\
& \quad\ \ 0.71177 = e^{-0.34}
\end{aligned}
$$

the result being correct to the last place, as the remainder-term has no influence.

52. To recapitulate, we must, in interpolating, be prepared for four kinds of errors:

1. Forcing-errors in the given table.
2. Grosser errors in the table.

3. Forcing-errors in each separate step of the calculation.
4. Errors due to neglecting the remainder-term.

The errors of type 2° are usually discovered in setting out the difference-table. The errors of type 3° are neutralized by performing the calculation with one or two decimals more than are required in the result. The errors of type 4° can only be dealt with by examining the remainder-term, and it is not sufficient—although many practical computers think so—to examine the difference-table without taking the nature of the function into account. If, for instance, we have the table

x	$f(x)$	Δ	Δ^2
0	100		
		10	
1	110		1
		11	
2	121		

it would be rash to think that, because the differences have a regular course and decrease rapidly, interpolation is safe, without taking account of the remainder-term. For, interpolating for $f(\frac{1}{2})$ to the second difference, we find $f(\frac{1}{2}) = 104.875$; but if it is known that the tabulated function is

$$f(x) = \frac{761}{3}x^3 - \frac{1521}{2}x^2 + \frac{3101}{6}x + 100,$$

it is seen that $f(\frac{1}{2}) = 200$. The remainder-term is, in fact,

$$R = \frac{1}{16}f^{(3)}(\xi) = 95.125.$$

53. If nothing whatever is known about the nature of the function, we are rather at a loss as to the problem of interpolation. The best thing to do in such cases is, when possible, to reduce the interval so much that linear interpolation becomes possible. It is true, that even linear interpolation is not legitimate without consideration of the remainder-term; but in many cases it is possible, by means of simple physical or statistical considerations, to assert, with a considerable degree of certainty, that the function is practically linear in a certain small interval. But no information about the behaviour of the differential coefficients of advanced order is obtainable in this way.

54. In certain cases where it is not possible to prove in a satisfactory way that the remainder-term is very small, it may yet be possible to assert that it is *probably* very small. For if we calculate a few divided differences of order $n + 1$ for arguments chosen *at random*, we may look upon these as samples of the order of magnitude of $\frac{1}{(n+1)!}f^{(n+1)}(\xi)$. If Δ^{n+1} is very small, this may, when the interval chosen as unit is sufficiently small, indicate that also $f^{(n+1)}$ is very small. But this is far from being a safe criterion, especially as there may be a connection or correlation between the choice of arguments and the properties of the function. Thus, if the values of the two functions

$$f(x) = \frac{1}{x+100}, \quad \varphi(x) = \frac{1}{x+100} + \sin \pi x,$$

are given for $x = 0, 1, 2, 3$ and 4, and we form the two difference-tables, these will be identical and the differences will decrease rapidly, although the remainder-terms are very different. As the remainder-term for f is very small, the interpolation for f will produce a good approximation, but not so in the case of φ which at many points, for instance $x = \frac{1}{2}$, differs greatly from f.

55. It is unavoidable that, in the applications, cases occur where one is forced to interpolate, possibly even with differences of a high order, without having any idea whatever as to the influence of the remainder-term. The result of an interpolation under such circumstances must be considered as an hypothesis, and not as a mathematically proved fact.

§6. Factorial Coefficients

56. In the theory of interpolation it is often necessary to expand a power of x in factorials, or a factorial in powers of x. The coefficients occurring in such expansions will be called *factorial coefficients*, and we shall, in this section, occupy ourselves with the means of tabulating them.

The expansion of x^r in descending factorials is of the form

$$x^r = \sum_0^r x^{(\nu)} \frac{\Delta^\nu 0^r}{\nu!}. \tag{1}$$

The numbers $\Delta^\nu 0^r$ are usually, with a not too happily chosen name, called *differences of nothing.* They are defined by

$$\Delta^m_{x=0} x^r = m^r - \binom{m}{1}(m-1)^r + \binom{m}{2}(m-2)^r - \dots$$

or

$$\Delta^m 0^r = \sum_0^m (-1)^\nu \binom{m}{\nu} (m-\nu)^r. \tag{2}$$

If, in §2 (16), we put $u_x = x$, $v_x = x^r$, we find

$$\Delta^n x^{r+1} = n\,\Delta^{n-1} x^r + (x+n)\,\Delta^n x^r$$

whence, putting $x = 0$, $n = m$,

$$\Delta^m 0^{r+1} = m\,\Delta^{m-1} 0^r + m\,\Delta^m 0^r$$

or, dividing by $m!$,

$$\frac{\Delta^m 0^{r+1}}{m!} = m\,\frac{\Delta^m 0^r}{m!} + \frac{\Delta^{m-1} 0^r}{(m-1)!}, \tag{3}$$

a relation which, together with the obvious values

$$\Delta\, 0^r = 1, \quad \frac{\Delta^m 0^m}{m!} = 1 \tag{4}$$

serves for the successive calculation of the coefficients in the expansion (1). These coefficients are, as appears from (3) and (4), all positive. The values of the first few of them are given in the following table.

	$\frac{\Delta}{1!}$	$\frac{\Delta^2}{2!}$	$\frac{\Delta^3}{3!}$	$\frac{\Delta^4}{4!}$	$\frac{\Delta^5}{5!}$	$\frac{\Delta^6}{6!}$	$\frac{\Delta^7}{7!}$	$\frac{\Delta^8}{8!}$	$\frac{\Delta^9}{9!}$	$\frac{\Delta^{10}}{10!}$
0^1	1									
0^2	1	1								
0^3	1	3	1							
0^4	1	7	6	1						
0^5	1	15	25	10	1					
0^6	1	31	90	65	15	1				
0^7	1	63	301	350	140	21	1			
0^8	1	127	966	1701	1050	266	28	1		
0^9	1	255	3025	7770	6951	2646	462	36	1	
0^{10}	1	511	9330	34105	42525	22827	5880	750	45	1

57. The same table may be used for determining the coefficients in the expansion of x^r in ascending factorials, or

$$x^r = \sum_0^r x^{(-\nu)} \frac{\nabla^\nu 0^r}{\nu!}. \tag{5}$$

For if, in (1), we put $x^{(\nu)} = (-1)^\nu (-x)^{(-\nu)}$ and replace, thereafter, x by $-x$, we find

$$x^r = \sum_{\nu=0}^r (-1)^{\nu+r} x^{(-\nu)} \frac{\Delta^\nu 0^r}{\nu!},$$

showing, by comparison with (5), that

$$\frac{\nabla^\nu 0^r}{\nu!} = (-1)^{\nu+r} \frac{\Delta^\nu 0^r}{\nu!}. \tag{6}$$

58. The development of x^r in central factorials

$$x^r = \sum_0^r x^{[\nu]} \frac{\delta^\nu 0^r}{\nu!} \tag{7}$$

leads to *central differences of nothing*, that is

$$\delta^m_{x=0} x^r = \left(\tfrac{m}{2}\right)^r - \binom{m}{1}\left(\tfrac{m}{2}-1\right)^r + \binom{m}{2}\left(\tfrac{m}{2}-2\right)^r - \ldots$$

or

$$\delta^m 0^r = \sum_0^m (-1)^\nu \binom{m}{\nu}\left(\tfrac{m}{2}-\nu\right)^r. \tag{8}$$

If, in this formula, we unite the first and last term, the second and last but one term, etc., we find that

$$\left.\begin{aligned} &\delta^m 0^r = 0 && (m + r \textit{ odd}) \\ &\delta^m 0^r = 2 \sum_{\nu=0}^{\nu<\frac{m}{2}} (-1)^\nu \binom{m}{\nu} \left(\frac{m}{2} - \nu\right)^r && (m + r \textit{ even}). \end{aligned}\right\} \quad (9)$$

It follows from these relations that (7) may be replaced by

$$\left.\begin{aligned} x^{2k} &= \sum_{\nu=1}^{k} x^{[2\nu]} \frac{\delta^{2\nu} 0^{2k}}{(2\nu)!} \\ x^{2k+1} &= \sum_{\nu=0}^{k} x^{[2\nu+1]} \frac{\delta^{2\nu+1} 0^{2k+1}}{(2\nu+1)!}. \end{aligned}\right\} \quad (10)$$

The central differences of nothing are all positive. For it is seen, that $\frac{1}{r!} \delta^m 0^r$ is the coefficient of x^r in the expansion of

$$\left(e^{\frac{x}{2}} - e^{-\frac{x}{2}}\right)^m = \sum_{0}^{m} (-1)^\nu \binom{m}{\nu} e^{\left(\frac{m}{2} - \nu\right)x},$$

and these coefficients are all positive, as

$$\left(e^{\frac{x}{2}} - e^{-\frac{x}{2}}\right)^m = \left[2 \sum_{\nu=0}^{\infty} \frac{1}{(2\nu+1)!} \left(\frac{x}{2}\right)^{2\nu+1}\right]^m.$$

The following tables of the coefficients of even, and of odd, order have been calculated by (9). They contain the coefficients of the first, and second, respectively, of equations (10). It is seen that the expansion of a power of x in central factorials contains a smaller number of terms than the expansion in descending or ascending factorials.

	$\frac{\delta^2}{2!}$	$\frac{\delta^4}{4!}$	$\frac{\delta^6}{6!}$	$\frac{\delta^8}{8!}$	$\frac{\delta^{10}}{10!}$
0^2	1				
0^4	1	1			
0^6	1	5	1		
0^8	1	21	14	1	
0^{10}	1	85	147	30	1

	$\frac{\delta}{1!}$	$\frac{\delta^3}{3!}$	$\frac{\delta^5}{5!}$	$\frac{\delta^7}{7!}$	$\frac{\delta^9}{9!}$
0^1	1				
0^3	$\frac{1}{4}$	1			
0^5	$\frac{1}{16}$	$\frac{5}{2}$	1		
0^7	$\frac{1}{64}$	$\frac{91}{16}$	$\frac{35}{4}$	1	
0^9	$\frac{1}{256}$	$\frac{205}{16}$	$\frac{483}{8}$	21	1

59. We now consider the expansion of the ascending factorial in powers of x

$$x^{(-r)} = \sum_{0}^{r} x^{\nu} \frac{D^{\nu} 0^{(-r)}}{\nu!}. \tag{11}$$

The numbers $D^{\nu} 0^{(-r)}$ are called *differential coefficients of nothing.* As the polynomial $x^{(-r-1)}$ is formed from $x^{(-r)}$ by multiplication by $x + r$, we have by Leibnitz' formula

$$\frac{D^m 0^{(-r-1)}}{m!} = r \frac{D^m 0^{(-r)}}{m!} + \frac{D^{m-1} 0^{(-r)}}{(m-1)!}, \tag{12}$$

a relation which, together with the easily proved relations

$$D0^{(-r)} = (r-1)!, \quad \frac{D^r 0^{(-r)}}{r!} = 1 \tag{13}$$

serves for the successive calculation of these coefficients. They are all positive, as appears from (12) and (13). The first few of them are given in the following table.

	$\frac{D}{1!}$	$\frac{D^2}{2!}$	$\frac{D^3}{3!}$	$\frac{D^4}{4!}$	$\frac{D^5}{5!}$	$\frac{D^6}{6!}$	$\frac{D^7}{7!}$	$\frac{D^8}{8!}$	$\frac{D^9}{9!}$	$\frac{D^{10}}{10!}$
$0^{(-1)}$	1									
$0^{(-2)}$	1	1								
$0^{(-3)}$	2	3	1							
$0^{(-4)}$	6	11	6	1						
$0^{(-5)}$	24	50	35	10	1					
$0^{(-6)}$	120	274	225	85	15	1				
$0^{(-7)}$	720	1764	1624	735	175	21	1			
$0^{(-8)}$	5040	13068	13132	6769	1960	322	28	1		
$0^{(-9)}$	40320	109584	118124	67284	22449	4536	546	36	1	
$0^{(-10)}$	362880	1026576	1172700	723680	269325	63273	9450	870	45	1

60. From the identity $x^{(r)} = (-1)^r (-x)^{(-r)}$ we derive

$$\frac{D^m 0^{(r)}}{m!} = (-1)^{m+r} \frac{D^m 0^{(-r)}}{m!}, \tag{14}$$

so that the same table may be used for determining the coefficients in the expansion

$$x^{(r)} = \sum_{0}^{r} x^{\nu} \frac{D^{\nu} 0^{(r)}}{\nu!}. \tag{15}$$

61. The development of the central factorial in powers of x

$$x^{[r]} = \sum_{0}^{r} x^{\nu} \frac{D^{\nu} 0^{[r]}}{\nu!} \tag{16}$$

can be split up into two parts, as the expansion of $x^{[2k]}$ contains only even powers of x, and the expansion of $x^{[2k+1]}$ only odd powers. Therefore, $D^m 0^{[r]} = 0$, if $m + r$ is an odd number. The two expansions are

$$\left.\begin{aligned} x^{[2k]} &= \sum_{\nu=1}^{k} x^{2\nu} \frac{D^{2\nu} 0^{[2k]}}{(2\nu)!} \\ x^{[2k+1]} &= \sum_{\nu=0}^{k} x^{2\nu+1} \frac{D^{2\nu+1} 0^{[2k+1]}}{(2\nu+1)!}. \end{aligned}\right\} \tag{17}$$

As the non-vanishing coefficients in the expansion of $x^{[r]}$ in powers of x have alternating signs, it is seen from (17) that $D^{2\nu} 0^{2k}$ and $D^{2\nu+1} 0^{2k+1}$ have both the sign $(-1)^{\nu+k}$.

The coefficients may be calculated by a recurrence formula. We first obtain from the identity $x^{[2k+2]} = (x^2 - k^2) x^{[2k]}$, by Leibnitz' formula,

$$D^{2\nu} x^{[2k+2]} =$$

$$(x^2-k^2) D^{2\nu} x^{[2k]} + \binom{2\nu}{1} D^{2\nu-1} x^{[2k]} \cdot D(x^2-k^2) + \binom{2\nu}{2} D^{2\nu-2} x^{[2k]} \cdot D^2(x^2-k^2)$$

$$= (x^2 - k^2) D^{2\nu} x^{[2k]} + 4\nu x D^{2\nu-1} x^{[2k]} + (2\nu)^{(2)} D^{2\nu-2} x^{[2k]}$$

consequently, if we put $x = 0$ and divide by $(2\nu)!$,

$$\frac{D^{2\nu} 0^{[2k+2]}}{(2\nu)!} = \frac{D^{2\nu-2} 0^{[2k]}}{(2\nu-2)!} - k^2 \frac{D^{2\nu} 0^{[2k]}}{(2\nu)!}; \tag{18}$$

the initial values being

$$\frac{D^2 0^{[2k]}}{2\,!} = (-1)^{k-1}[(k-1)\,!]^2; \frac{D^{2k}\,0^{[2k]}}{(2k)\,!} = 1. \tag{19}$$

In a similar way, from the identity

$$x^{[2k+1]} = \left(x^2 - \frac{(2k-1)^2}{4}\right) x^{[2k-1]}$$

we obtain the recurrence formula

$$\frac{D^{2\nu+1}\,0^{[2k+1]}}{(2\nu+1)\,!} = \frac{D^{2\nu-1}\,0^{[2k-1]}}{(2\nu-1)\,!} - \frac{(2k-1)^2}{4} \frac{D^{2\nu+1}\,0^{[2k-1]}}{(2\nu+1)\,!}; \tag{20}$$

the initial values being

$$D0^{[2k-1]} = (-1)^{k-1}\left(\frac{1.3\ldots(2k-3)}{2^{k-1}}\right)^2; \frac{D^{2k-1}\,0^{[2k-1]}}{(2k-1)\,!} = 1. \tag{21}$$

By means of these formulas the numbers in the following tables have been calculated.

	$\frac{D^2}{2!}$	$\frac{D^4}{4!}$	$\frac{D^6}{6!}$	$\frac{D^8}{8!}$	$\frac{D^{10}}{10!}$
$0^{[2]}$	1				
$0^{[4]}$	−1	1			
$0^{[6]}$	4	−5	1		
$0^{[8]}$	−36	49	−14	1	
$0^{[10]}$	576	−820	273	−30	1

	$\frac{D}{1!}$	$\frac{D^3}{3!}$	$\frac{D^5}{5!}$	$\frac{D^7}{7!}$	$\frac{D^9}{9!}$
$0^{[1]}$	1				
$0^{[3]}$	$-\frac{1}{4}$	1			
$0^{[5]}$	$\frac{9}{16}$	$-\frac{5}{2}$	1		
$0^{[7]}$	$-\frac{225}{64}$	$\frac{259}{16}$	$-\frac{35}{4}$	1	
$0^{[9]}$	$\frac{11025}{256}$	$-\frac{3229}{16}$	$\frac{987}{8}$	−21	1

62. As a control on the calculation of the tables in this section we may employ the following relations. From (11) we obtain for $x = 1$

$$\sum_{0}^{r} \frac{D^{\nu}0^{(-r)}}{\nu\,!} = r\,! \tag{22}$$

and from (17) for $x = 1$

$$\left.\begin{aligned} \sum_{\nu=1}^{k} \frac{D^{2\nu}\, 0^{[2k]}}{(2\nu)\,!} &= 0 \qquad (k > 1) \\ \sum_{\nu=0}^{k} \frac{D^{2\nu+1}\, 0^{[2k+1]}}{(2\nu+1)\,!} &= (-1)^{k-1} \frac{3}{2^{2k}} (9-4)(25-4)\ldots[(2k-1)^2-4] \quad (k > 1). \end{aligned}\right\} \quad (23)$$

Further, we obtain from (1), assuming $r > 1$, dividing by x and putting $x = 0$,

$$\sum_{\nu=1}^{r} (-1)^{\nu-1} \frac{\Delta^{\nu} 0^{r}}{\nu} = 0 \qquad (r > 1) \qquad (24)$$

and from (10), by a similar process,

$$\left.\begin{aligned} \sum_{\nu=1}^{k} (-1)^{\nu-1} [(\nu-1)\,!]^2 \frac{\delta^{2\nu} 0^{2k}}{(2\nu)\,!} &= 0 \qquad (k > 1) \\ \sum_{\nu=0}^{k} (-1)^{\nu} \frac{[1.3\ \ldots\ (2\nu-1)]^2}{2^{2\nu}} \frac{\delta^{2\nu+1}\, 0^{2k+1}}{(2\nu+1)\,!} &= 0 \qquad (k > 0). \end{aligned}\right\} \quad (25)$$

§7. Numerical Differentiation

63. The problem of calculating, by means of a table of a function, the successive differential coefficients of the function, is called *numerical differentiation*. It is solved by expressing the required differential coefficient in terms of the successive differences of the function.

As regards the first differential coefficient, the problem is immediately solved by the interpolation-formulas of §4. Thus, from the formula with descending differences, we get

$$\frac{f(x) - f(0)}{x} = \sum_{1}^{n} \frac{(x-1)^{(\nu-1)}}{\nu\,!} \Delta^{\nu} f(0) + \frac{(x-1)^{(n)}}{(n+1)\,!} f^{(n+1)}(\xi),$$

and from this, for $x \to 0$,

$$f'(0) = \sum_{1}^{n} \frac{(-1)^{\nu-1}}{\nu} \Delta^{\nu} f(0) + \frac{(-1)^{n}}{n+1} f^{(n+1)}(\xi),$$

consequently, if everywhere, instead of $f(t)$, we write $f(t + x)$,

$$f'(x) = \sum_{1}^{n} \frac{(-1)^{\nu-1}}{\nu} \Delta^{\nu} f(x) + \frac{(-1)^n}{n+1} f^{(n+1)}(\xi), \tag{1}$$

x being one of the arguments of the table.

In a similar way we obtain from the formula with ascending differences

$$f'(x) = \sum_{1}^{n} \frac{1}{\nu} \nabla^{\nu} f(x) + \frac{1}{n+1} f^{(n+1)}(\xi) \tag{2}$$

and from Stirling's formula

$$f'(x) = \sum_{0}^{r-1} (-1)^{\nu} \frac{(\nu!)^2}{(2\nu+1)!} \square\, \delta^{2\nu+1} f(x) + (-1)^r \frac{(r!)^2}{(2r+1)!} f^{(2r+1)}(\xi); \tag{3}$$

finally from the interpolation-formula with central differences, §4 (33),

$$\left.\begin{aligned} f'(x+\tfrac{1}{2}) &= \sum_{0}^{r-1} (-1)^{\nu} \frac{[1.3\ldots(2\nu-1)]^2}{2^{2\nu}(2\nu+1)!} \delta^{2\nu+1} f(x+\tfrac{1}{2}) + R \\ R &= (-1)^r \frac{[1.3\ldots(2r-1)]^2}{2^{2r}(2r+1)!} f^{(2r+1)}(\xi). \end{aligned}\right\} \tag{4}$$

A formula with arbitrary arguments is obtained if, in §4 (8), we transfer $f(a_0)$ to the left-hand side, divide by a_0 and let $a_0 \to 0$. The result is

$$\left.\begin{aligned} f'(0) &= f(0, a_1) - a_1 f(0, a_1, a_2) + a_1 a_2 f(0, a_1, a_2, a_3) - \ldots \\ &+(-1)^{n-1} a_1 a_2 \ldots a_{n-1} f(0, a_1, \ldots a_n) + (-1)^n a_1 a_2 \ldots a_n \frac{f^{(n+1)}(\xi)}{(n+1)!} \end{aligned}\right\} \tag{5}$$

In this formula, $a_1, a_2, \ldots a_n$ denote the arguments measured from that point (the origin) at which the differential coefficient is required.

64. The first few terms of the formulas (1) – (4) are, for practical purposes, written

$$D = \Delta - \frac{\Delta^2}{2} + \frac{\Delta^3}{3} - \frac{\Delta^4}{4} + \ldots \tag{6}$$

$$D = \nabla + \frac{\nabla^2}{2} + \frac{\nabla^3}{3} + \frac{\nabla^4}{4} + \ldots \tag{7}$$

$$D = \Box\delta - \frac{1}{6}\Box\delta^3 + \frac{1}{30}\Box\delta^5 - \frac{1}{140}\Box\delta^7 + \ldots \tag{8}$$

$$D = \delta - \frac{1}{24}\delta^3 + \frac{3}{640}\delta^5 - \frac{5}{7168}\delta^7 + \ldots \tag{9}$$

In the first three of these formulas, $f(x)$ has been left out on both sides; in the last, $f(x + \frac{1}{2})$. The practical application is so simple that it seems superfluous to give any numerical examples. The coefficients, being also coefficients in the remainder-term, show that the two last of these formulas, particularly the last one, are much to be preferred to the others, which, therefore, should only be used at the beginning and end of a table.

65. If differential coefficients of higher order than the first are wanted, we may calculate a section of a table of $f'(x)$, large enough for calculating a similar table of $f''(x)$, by the same formula; and in this way we may continue; the accuracy obtained being, at each step, controlled by the remainder-term.

66. But we may also form a direct expression for the differential coefficient of an arbitrary order. Let us, for abbreviation, write

$$\left.\begin{aligned} x_\nu &= (x - a_0)(x - a_1)\ldots(x - a_{\nu-1}),\, x_0 = 1; \\ t_\nu &= (t - a_0)(t - a_1)\ldots(t - a_{\nu-1}),\, t_0 = 1. \end{aligned}\right\} \tag{10}$$

Newton's formula with divided differences is, then, written in the form

$$f(x) = \sum_{\nu=0}^{n} x_\nu f(a_0, \ldots a_\nu) + x_{n+1} f(x, a_0, \ldots a_n), \tag{11}$$

so that

$$f^{(m)}(x) = \sum_{\nu=m}^{n} f(a_0, \ldots a_\nu)\, D^m x_\nu + D^m [x_{n+1} f(x, a_0, \ldots a_n)]. \tag{12}$$

We now consider the function

$$\varphi(x) \equiv x_{n+1} f(x, a_0, \ldots a_n) - K \frac{x_{n+1}}{(n+1)!}, \tag{13}$$

K denoting a constant. Assuming for the time being that all the a_ν are different, it is clear, on account of the factor x_{n+1}, that $\varphi(x)$ vanishes at least $n + 1$ times. If we differentiate (13) m $(\leq n)$ times, it is seen by repeated application of Rolle's theorem that the function

$$\varphi^{(m)}(x) = D^m [x_{n+1} f(x, a_0, \ldots a_n)] - K \frac{D^m x_{n+1}}{(n+1)!} \tag{14}$$

vanishes at least $n - m + 1$ times in the interior of an interval limited by the smallest and the largest of the numbers $a_0, a_1, \ldots a_n$.

Now, let t be a number, independent of x, and not belonging to the interior of the said interval (while t may be one of the limiting points of the interval). It follows, in particular, that $D^m t_{n+1} \neq 0$; for $D^m x_{n+1}$ has, according to Rolle's theorem, exactly $n - m + 1$ roots which are all situated in the interior of the interval. It is, therefore, always possible to determine a particular value of K, such that

$$D^m [t_{n+1} f(t, a_0, \ldots a_n)] - K \frac{D^m t_{n+1}}{(n+1)!} = 0$$

or $\varphi^{(m)}(t) = 0$. If K has this value, $\varphi^{(m)}(x)$ must vanish at least $n - m + 2$ times, as the new root t has been added which must be different from the $n - m + 1$ others, since these all belong to the interior of the interval. But from this follows by Rolle's theorem, if we differentiate (14) $n - m + 1$ times, that the function $\varphi^{(n+1)}(x)$ which may, by (11), be written

$$\varphi^{(n+1)}(x) = f^{(n+1)}(x) - K$$

vanishes at least once in the interior of an interval, limited by the smallest and the largest of the numbers $t, a_0, a_1, \ldots a_n$. Within this interval there exists, therefore, a number ξ, such that $K = f^{(n+1)}(\xi)$. We have, thus, proved the general relation

$$D^m [t_{n+1} f(t, a_0, \ldots a_n)] = f^{(n+1)}(\xi) \frac{D^m t_{n+1}}{(n+1)!} \tag{15}$$

on the assumptions made as regards the position of t.

By (15) we obtain, finally, from (12)

$$f^{(m)}(x) = \sum_{\nu=m}^{n} f(a_0, \ldots a_\nu) D^m x_\nu + f^{(n+1)}(\xi) \frac{D^m x_{n+1}}{(n+1)!} \qquad (16)$$

which is valid, like (15), provided only the variable *does not belong to the interior of the interval limited by the smallest and the largest of the numbers* $a_0, a_1, \ldots a_n$.

It was assumed, for the time being, that all the a_ν are different; but this assumption may now be discarded, as all the terms in (16) retain a meaning, if two or more of the numbers a_ν coincide.

67. (16) expresses that Newton's formula with divided differences may be differentiated $m(\leq n)$ times, *exactly as if* $f^{(n+1)}(\xi)$ *were a constant*, provided the resulting formula is not used for values of the variable, situated between the smallest and the largest of the numbers $a_0, a_1, \ldots a_n$. The same property appertains to formulas which—as Stirling's formula, stopping at a difference of even order, and Bessel's formula, stopping at a difference of odd order—are arithmetical means of particular cases of Newton's formula with the same remainder-term.

68. As a special case of (16) we obtain for $x = 0$, $a_\nu = \nu$

$$f^{(m)}(0) = \sum_{\nu=m}^{n} \Delta^\nu f(0) \frac{D^m 0^{(\nu)}}{\nu!} + f^{(n+1)}(\xi) \frac{D^m 0^{(n+1)}}{(n+1)!}$$

or, replacing $f(t)$ by $f(t+x)$, Markoff's formula[1]

$$f^{(m)}(x) = \sum_{\nu=m}^{n} \Delta^\nu f(x) \frac{D^m 0^{(\nu)}}{\nu!} + f^{(n+1)}(\xi) \frac{D^m 0^{(n+1)}}{(n+1)!} \qquad (17)$$

where $m \leq n$ and $x \leq \xi < x + n$.

The first term in this formula is, as might have been anticipated, $\Delta^m f(x)$. The case $n = m$ is of interest, as it shows with what degree of approximation $\Delta^m f(x)$ and $f^{(m)}(x)$ may replace each other. We find

$$\Delta^m f(x) = f^{(m)}(x) + \frac{m}{2} f^{(m+1)}(\xi). \qquad (18)$$

[1] Markoff: Differenzenrechnung, p. 21.

The formula with ascending differences, corresponding to (17), is obtained by replacing Δ by ∇ and $D^m\, 0^{(\nu)}$ by $D^m\, 0^{(-\nu)}$. The result is

$$f^{(m)}(x) = \sum_{\nu=m}^{n} \nabla^\nu f(x) \frac{D^m\, 0^{(-\nu)}}{\nu\,!} + f^{(n+1)}(\xi) \frac{D^m\, 0^{(-n-1)}}{(n+1)\,!} \tag{19}$$

69. In (16), it is only permissible to put $x = 0$, if all the a_ν have the same sign. In order to obtain a formula with central differences, corresponding to (17), we must, therefore, go back to (12). Applying Leibnitz' theorem to the remainder-term, we write this formula

$$f^{(m)}(x) = \sum_{\nu=m}^{n} f(a_0, \ldots a_\nu)\, D^m x_\nu + R, \tag{20}$$

$$R = \sum_{\nu=0}^{m} \binom{m}{\nu} D^\nu x_{n+1} \,.\, D^{m-\nu} f(x, a_0, \ldots a_n). \tag{21}$$

Now

$$D^{m-\nu} f(x, a_0, \ldots a_n) = (m-\nu)\,!\, f(x, \ldots x, a_0, \ldots a_n),$$

the argument x being repeated $m - \nu + 1$ times; hence

$$D^{m-\nu} f(x, a_0, \ldots a_n) = \frac{(m-\nu)\,!}{(m+n-\nu+1)\,!} f^{(m+n-\nu+1)}(\xi),$$

so that, by inserting this value in (21), we get

$$R = \sum_{\nu=0}^{m} \frac{D^\nu x_{n+1}}{\nu\,!\,(m+n-\nu+1)^{(n-\nu+1)}} f^{(m+n-\nu+1)}(\xi) \tag{22}$$

where, of course, ξ need not have the same value in the different differential coefficients.

Formula (20) with the remainder-term (22) is valid for all x; on the other hand the remainder-term is not so simple as that of (16); yet the complication is more apparent than real, and it is often possible, by means of (22), to find limits to the error involved.

70. We find, for instance, differentiating Stirling's formula an even number of times, putting, after the differentiation, $x = 0$, and remembering that $D^{2s}\,0^{[2\nu]-1} = 0$,

$$\left.\begin{aligned} f^{(2s)}(0) &= \sum_{\nu=s}^{r-1} \frac{D^{2s}\,0^{[2\nu]}}{(2\nu)!}\,\delta^{2\nu} f(0) + R \\ R &= \sum_{\nu=0}^{2s} \frac{D^{\nu}\,0^{[2r]-1}}{\nu!\,(2s+2r-\nu-1)^{(2r-\nu-1)}} f^{(2s+2r-\nu-1)}(\xi). \end{aligned}\right\}$$

If now we note that $D^{\nu}\,0^{[n]-1} = \dfrac{1}{\nu+1} D^{\nu+1}\,0^{[n]}$, as is seen by expanding $x^{[n]}$ in powers of x and dividing by x; and if we replace $f(t)$ by $f(t+x)$, we finally obtain

$$\left.\begin{aligned} f^{(2s)}(x) &= \sum_{\nu=s}^{r-1} \frac{D^{2s}\,0^{[2\nu]}}{(2\nu)!}\,\delta^{2\nu} f(x) + R \\ R &= \sum_{\nu=1}^{s} \frac{D^{2\nu}\,0^{[2r]}}{(2\nu)!\,(2s+2r-2\nu)^{(2r-2\nu)}} f^{(2s+2r-2\nu)}(\xi). \end{aligned}\right\} \quad (23)$$

The formula for the second differential coefficient is particularly simple; we have

$$f''(x) = \sum_{\nu=1}^{r-1} \frac{D^2\,0^{[2\nu]}}{(2\nu)!}\,\delta^{2\nu} f(x) + \frac{D^2\,0^{[2r]}}{(2r)!} f^{(2r)}(\xi), \quad (24)$$

or, by §6 (19),

$$\left.\begin{aligned} &f''(x) = \\ &\sum_{\nu=1}^{r-1} (-1)^{\nu-1} \frac{2[(\nu-1)!]^2}{(2\nu)!}\,\delta^{2\nu} f(x) + (-1)^{r-1} \frac{2[(r-1)!]^2}{(2r)!} f^{(2r)}(\xi). \end{aligned}\right\} \quad (25)$$

71. If we differentiate Stirling's formula an odd number of times and put, thereafter, $x = 0$, we find first

$$\left.\begin{aligned} f^{(2s-1)}(0) &= \sum_{\nu=s}^{r-1} \frac{D^{2s-1}\,0^{[2\nu]-1}}{(2\nu-1)!} \,\square\, \delta^{2\nu-1} f(0) + R \\ R &= \sum_{\nu=0}^{2s-1} \frac{D^{\nu}0^{[2r]-1}}{\nu!\,(2s+2r-\nu-2)^{(2r-\nu-1)}} f^{(2s+2r-\nu-2)}(\xi) \end{aligned}\right\}$$

and from this, in a similar way,

$$\left.\begin{aligned} f^{(2s-1)}(x) &= \sum_{\nu=s}^{r-1} \frac{D^{2s}0^{[2\nu]}}{(2\nu-1)!\,2s} \Box\, \delta^{2\nu-1} f(x) + R \\ R &= \sum_{\nu=1}^{s} \frac{D^{2\nu}\,0^{[2r]}}{(2\nu)!\,(2s+2r-2\nu-1)^{(2r-2\nu)}} f^{(2s+2r-2\nu-1)}(\xi). \end{aligned}\right\} \quad (26)$$

For $s = 1$, we have again (3).

72. An analogous set of formulas is obtained from Bessel's formula. Thus, differentiating this formula an even number of times, we find for $x = 0$

$$\left.\begin{aligned} f^{(2s)}\left(\tfrac{1}{2}\right) &= \sum_{\nu=s}^{r-1} \frac{D^{2s}\,0^{[2\nu+1]-1}}{(2\nu)!} \Box\, \delta^{2\nu} f\left(\tfrac{1}{2}\right) + R \\ R &= \sum_{\nu=0}^{2s} \frac{D^{\nu}\,0^{[2r+1]-1}}{\nu!\,(2s+2r-\nu)^{(2r-\nu)}} f^{(2s+2r-\nu)}(\xi) \end{aligned}\right\}$$

and from this, as above,

$$\left.\begin{aligned} f^{(2s)}\left(x+\tfrac{1}{2}\right) &= \sum_{\nu=s}^{r-1} \frac{D^{2s+1}\,0^{[2\nu+1]}}{(2\nu)!\,(2s+1)} \Box\, \delta^{2\nu} f\left(x+\tfrac{1}{2}\right) + R \\ R &= \sum_{\nu=0}^{s} \frac{D^{2\nu+1}\,0^{[2r+1]}}{(2\nu+1)!\,(2s+2r-2\nu)^{(2r-2\nu)}} f^{(2s+2r-2\nu)}(\xi). \end{aligned}\right\} \quad (27)$$

For $s = 1$, we get

$$f''\left(x+\tfrac{1}{2}\right) =$$

$$\sum_{\nu=1}^{r-1} \frac{D^3\,0^{[2\nu+1]}}{3\,(2\nu)!} \Box\, \delta^{2\nu} f\left(x+\tfrac{1}{2}\right) + \frac{D^3\,0^{[2r+1]}}{3\,(2r)!} f^{(2r)}(\xi) + \frac{2\,D\,0^{[2r+1]}}{(2r+2)!} f^{(2r+2)}(\xi). \quad (28)$$

73. If, on the other hand, Bessel's formula is differentiated an odd number of times, we find, for $x = 0$,

$$\left.\begin{aligned} f^{(2s+1)}\left(\tfrac{1}{2}\right) &= \sum_{\nu=s}^{r-1} \frac{D^{2s+1}\,0^{[2\nu+1]}}{(2\nu+1)!} \delta^{2\nu+1} f\left(\tfrac{1}{2}\right) + R \\ R &= \sum_{\nu=0}^{2s+1} \frac{D^{\nu}\,0^{[2r+1]-1}}{\nu!\,(2s+2r-\nu+1)^{(2r-\nu)}} f^{(2s+2r-\nu+1)}(\xi) \end{aligned}\right\} \quad (29)$$

and from this, as above,

$$\left.\begin{aligned} f^{(2s+1)}(x+\tfrac{1}{2}) &= \sum_{\nu=s}^{r-1} \frac{D^{2s+1}\,0^{[2\nu+1]}}{(2\nu+1)!}\,\delta^{2\nu+1} f(x+\tfrac{1}{2}) + R \\ R &= \sum_{\nu=0}^{s} \frac{D^{2\nu+1}\,0^{[2r+1]}}{(2\nu+1)!\,(2s+2r-2\nu+1)^{(2r-2\nu)}}\, f^{(2s+2r-2\nu+1)}(\xi). \end{aligned}\right\} \quad (30)$$

For $s = 0$, we have again (4).

74. For practical purposes the formulas for the second differential coefficient by (17), (19), (24) and (28) are written

$$D^2 = \Delta^2 - \Delta^3 + \frac{11}{12}\Delta^4 - \frac{5}{6}\Delta^5 + \frac{137}{180}\Delta^6 - \ldots\ldots \quad (31)$$

$$D^2 = \nabla^2 + \nabla^3 + \frac{11}{12}\nabla^4 + \frac{5}{6}\nabla^5 + \frac{137}{180}\nabla^6 + \ldots\ldots \quad (32)$$

$$D^2 = \delta^2 - \frac{\delta^4}{12} + \frac{\delta^6}{90} - \frac{\delta^8}{560} + \ldots \quad (33)$$

$$D^2 = \square\,\delta^2 - \frac{5}{24}\square\,\delta^4 + \frac{259}{5760}\square\,\delta^6 - \frac{3229}{322560}\square\,\delta^8 + \ldots \quad (34)$$

In the first three of these formulas the factor $f(x)$ has been left out; in the last, the factor is $f(x + \frac{1}{2})$. The coefficients are, as far as the three first formulas are concerned, also coefficients of the remainder-term. The two first formulas are not of much practical value, and are only applied where central differences cannot be had.

75. The Error-Test is immediately applicable to several of the above formulas. Confining ourselves to the case where the first neglected term does not vanish, the conditions for the applicability are: in the case of (1) and (17), that $f^{(n+1)}$ and $f^{(n+2)}$ keep their signs and have the same sign; in the case of (2) and (19), that $f^{(n+1)}$ and $f^{(n+2)}$ keep their signs and have opposite signs; in the case of (3) and (4), that $f^{(2r+1)}$ and $f^{(2r+3)}$ keep their signs and have the same sign; in the case of (24), that $f^{(2r)}$ and $f^{(2r+2)}$ keep their signs and have the same sign.

76. A problem related to numerical differentiation is the problem of expressing a difference of a given order by means of the successive differential coefficients of the function. As regards the first difference, the problem is solved by Taylor's formula, and no further comments are necessary. As regards the differences of higher order, a formula of a similar nature to (16) may be derived by the following considerations.

Let $\varphi(x)$ be defined by

$$\varphi(x) \equiv f(x) - \sum_{0}^{n} \frac{x^\nu}{\nu!} f^{(\nu)}(0) - K \frac{x^{n+1}}{(n+1)!}, \tag{35}$$

K denoting a constant. For this function $\varphi(x)$ we have

$$\varphi^{(\nu)}(0) = 0 \qquad (\nu = 0, 1, 2, \ldots n), \tag{36}$$

$$\varphi^{(n+1)}(x) = f^{(n+1)}(x) - K. \tag{37}$$

Therefore, by Maclaurin's formula, assuming $m \leq n$,

$$\left.\begin{aligned} \varphi^{(m)}(x) &= \sum_{\nu=0}^{n-m} \frac{x^\nu}{\nu!} \varphi^{(m+\nu)}(0) + \frac{x^{n-m+1}}{(n-m+1)!} \varphi^{(n+1)}(\theta) \\ &= \frac{x^{n-m+1}}{(n-m+1)!} \varphi^{(n+1)}(\theta) \qquad (0 < \theta < x). \end{aligned}\right\} \tag{38}$$

Now, let $\varphi(a_0, a_1, \ldots a_m)$ and $(a_0, a_1, \ldots a_m)^\nu$ denote the mth divided differences of $\varphi(x)$ and x^ν respectively, formed with the arguments $a_0, a_1, \ldots a_m$. As $(a_0, a_1, \ldots a_m)^\nu = 0$ for $\nu < m$, we obtain from (35)

$$\varphi(a_0, \ldots a_m) = f(a_0, \ldots a_m) - \sum_{m}^{n} \frac{(a_0, \ldots a_m)^\nu}{\nu!} f^{(\nu)}(0) - K \frac{(a_0, \ldots a_m)^{n+1}}{(n+1)!}. \tag{39}$$

We assume, next, that *all the a_ν have the same sign* (or vanish); neglecting the trivial case where the a_ν are all zero, we may, for instance, assume that

$$a_\nu \geq 0 \qquad (\nu = 0, 1, \ldots m), \tag{40}$$

as the proof may be repeated line for line in the case where the a_ν are negative instead of positive.

As we have

$$(a_0, \ldots a_m)^{n+1} = \frac{D^m \xi^{n+1}}{m!} = \binom{n+1}{m} \xi^{n-m+1}, \tag{41}$$

this expression must, then, be positive, ξ being situated between the smallest and the largest of the numbers a_ν. It is, therefore, possible to find a number K, such that the equation

$$f(a_0, \ldots a_m) = \sum_m^n \frac{(a_0, \ldots a_m)^\nu}{\nu!} f^{(\nu)}(0) + K \frac{(a_0, \ldots a_m)^{n+1}}{(n+1)!} \tag{42}$$

is satisfied and, consequently,

$$\varphi(a_0, \ldots a_m) = 0. \tag{43}$$

On the other hand we have

$$\varphi(a_0, \ldots a_m) = \frac{\varphi^{(m)}(\xi)}{m!}, \tag{44}$$

the new ξ being also positive. Therefore, by (38) and (37),

$$\varphi(a_o, \ldots a_m) = \frac{\xi^{n-m+1}}{m!\,(n-m+1)!} \varphi^{(n+1)}(\theta)$$

$$= \frac{\xi^{n-m+1}}{m!\,(n-m+1)!} [f^{(n+1)}(\theta) - K]$$

where $0 < \theta < \xi$. But from this follows, ξ being positive, that $\varphi(a_o, \ldots a_m)$ can only vanish, if $K = f^{(n+1)}(\theta)$. Inserting this value in (42), we find finally, writing ξ instead of θ,

$$f(a_o, \ldots a_m) = \sum_m^n f^{(\nu)}(0) \frac{(a_o, \ldots a_m)^\nu}{\nu!} + f^{(n+1)}(\xi) \frac{(a_o, \ldots a_m)^{n+1}}{(n+1)!} \tag{45}$$

on the assumption that *all the* a_ν *have the same sign* (or vanish).

77. As an application, we will take $a_\nu = \nu$. We obtain then, multiplying by $m!$ and replacing $f(t)$ by $f(t + x)$, the following formula which is also due to Markoff (l.c., p. 23):

$$\Delta^m f(x) = \sum_{\nu=m}^{n} f^{(\nu)}(x) \frac{\Delta^m 0^\nu}{\nu!} + f^{(n+1)}(\xi) \frac{\Delta^m 0^{n+1}}{(n+1)!}. \tag{46}$$

The application of the Error-Test to this formula leads to the result that the error is numerically smaller than the first neglected term and has the same sign, if this term does not vanish, and if $f^{(n+1)}$ and $f^{(n+2)}$ keep their signs and have opposite signs.

78. The corresponding formula for $\delta^m f(x)$ is not obtained by this method, as we have assumed that all the a_ν have the same sign. It can be obtained from (11), performing δ^m on both sides and putting, thereafter, $x = a_0 = a_1 = \ldots = a_n = 0$. The remainder-term is first transformed by §2 (18), whereafter the unknown differences are replaced by differential coefficients with unknown arguments ξ. We shall not, however, go into details, as these formulas are not of much importance from a practical point of view.

§ 8. Construction of Tables

79. One of the most important applications of the theory of interpolation is its application to the construction of tables. If the function to be tabulated is a polynomial of low degree, the application is immediate. If, for instance, we want to tabulate the function $f(x) = 3x - x^3$ for intervals of 0.01, we may proceed as shown in the following section of the calculation.

x	$3x - x^3$	Δ	Δ^2	Δ^3
0.00	0.000000			
		29999		
0.01	0.029999		−6	
		29993		−6
0.02	0.059992		−12	
		29981		−6
0.03	0.089973		−18	
		29963		−6
0.04	0.119936		−24	
		29939		−6
0.05	0.149875		−30	
		29909		−6
0.06	0.179784		−36	
		29873		
0.07	0.209657			
.	.	.	.	.
.	.	.	.	.
.	.	.	.	.

The third difference being constant, it is only necessary to calculate the 4 first values of the function directly; these have, in the table, been indicated by a frame. All the remainder of the calculation consists of subtractions and additions. We first form all the differences (of the first, second and third order) which can be formed by the 4 directly calculated values; then the column headed Δ^3 may be filled up, as this difference is constant; thereafter, in succession, the columns Δ^2, Δ and $f(x)$. As a control, we may beforehand calculate directly a suitable number of values of $f(x)$, e.g., every tenth.

80. In more complicated cases we calculate directly a number of equidistant values of the function, for a suitable interval, and, thereafter, subdivide the interval by means of a method, due, in principle, to Briggs.[1]

The first step is, by means of the given differences of various orders, corresponding to the larger interval, to calculate the corresponding differences for the smaller interval.

We have by Stirling's formula, leaving aside the remainder-term

$$\frac{E^x - E^{-x}}{2x} = \Box\,\delta + \frac{x^2 - 1}{3\,!}\,\Box\,\delta^3 + \frac{(x^2-1)\,(x^2-4)}{5\,!}\,\Box\,\delta^5 \ldots \quad (1)$$

and by Bessel's formula

$$\frac{E^x - E^{-x}}{2x} = \delta + \frac{x^2 - \frac{1}{4}}{3\,!}\,\delta^3 + \frac{(x^2 - \frac{1}{4})\,(x^2 - \frac{9}{4})}{5\,!}\,\delta^5 + \ldots \quad (2)$$

If we introduce the notation

$$\delta_h = E^{\frac{h}{2}} - E^{-\frac{h}{2}}, \quad (3)$$

we therefore have, by (1), if for x we take a positive integer n,

$$\frac{1}{2n}\,\delta_{2n} = \Box\,\delta + \frac{n^2 - 1}{3\,!}\,\Box\,\delta^3 + \frac{(n^2-1)\,(n^2-4)}{5\,!}\,\Box\,\delta^5 + \ldots, \quad (4)$$

the expansion containing only a finite number of terms, so that no consideration of the remainder-term is needed, if all the terms in (4) are used.

If, in (2), we put $x = \frac{\mu}{2}$, μ denoting an *odd* positive number, we obtain the following expansion which is also finite

$$\frac{1}{\mu}\,\delta_\mu = \delta + \frac{\mu^2 - 1}{3\,!\,2^2}\,\delta^3 + \frac{(\mu^2-1)\;(\mu^2-9)}{5\,!\,2^4}\,\delta^5 + \ldots (\mu\ \textit{odd}). \quad (5)$$

[1] Thiele: Interpolationsrechnung §28.

Repetitions of the operation $\frac{1}{r}\,\delta_r$ may be expressed by multiplying (4) or (5) several times by itself. In the case of (4), all even powers of $\square$ may be removed by means of the relation $\square^2 = 1 + \frac{\delta^2}{4}$, so that the result at most contains the first power of $\square$, while everything else is expressed by δ. The expansions obtained in this way are, of course, also finite.

81. We obtain, for instance, from (4) for $n = 1$, leaving out differences of higher order than the 8th,

$$\left.\begin{aligned}
\tfrac{1}{2}\,\delta_2 &= \square\,\delta \\
(\tfrac{1}{2}\,\delta_2)^2 &= \delta^2 + \tfrac{1}{4}\,\delta^4 \\
(\tfrac{1}{2}\,\delta_2)^3 &= \square\,\delta^3 + \tfrac{1}{4}\,\square\,\delta^5 \\
(\tfrac{1}{2}\,\delta_2)^4 &= \delta^4 + \tfrac{1}{2}\,\delta^6 + \tfrac{1}{16}\,\delta^8 \\
(\tfrac{1}{2}\,\delta_2)^5 &= \square\,\delta^5 + \tfrac{1}{2}\,\square\,\delta^7 + \ldots \\
(\tfrac{1}{2}\,\delta_2)^6 &= \delta^6 + \tfrac{3}{4}\,\delta^8 + \ldots \\
(\tfrac{1}{2}\,\delta_2)^7 &= \square\,\delta^7 + \ldots \\
(\tfrac{1}{2}\,\delta_2)^8 &= \delta^8 + \ldots
\end{aligned}\right\}\quad(6)$$

further, from (5), putting $\mu = 3$,

$$\left.\begin{aligned}
\tfrac{1}{3}\,\delta_3 &= \delta + \tfrac{1}{3}\,\delta^3 \\
(\tfrac{1}{3}\,\delta_3)^2 &= \delta^2 + \tfrac{2}{3}\,\delta^4 + \tfrac{1}{9}\,\delta^6 \\
(\tfrac{1}{3}\,\delta_3)^3 &= \delta^3 + \delta^5 + \tfrac{1}{3}\,\delta^7 + \ldots \\
(\tfrac{1}{3}\,\delta_3)^4 &= \delta^4 + \tfrac{4}{3}\,\delta^6 + \tfrac{2}{3}\,\delta^8 + \ldots \\
(\tfrac{1}{3}\,\delta_3)^5 &= \delta^5 + \tfrac{5}{3}\,\delta^7 + \ldots \\
(\tfrac{1}{3}\,\delta_3)^6 &= \delta^6 + 2\delta^8 + \ldots \\
(\tfrac{1}{3}\,\delta_3)^7 &= \delta^7 + \ldots \\
(\tfrac{1}{3}\,\delta_3)^8 &= \delta^8 + \ldots
\end{aligned}\right\}\quad(7)$$

and for $\mu = 5$

$$\left.\begin{aligned}
0.2\delta_5 &= \delta + \delta^3 + 0.2\delta^5 \\
(0.2\delta_5)^2 &= \delta^2 + 2\delta^4 + 1.4\delta^6 + 0.4\delta^8 + \ldots \\
(0.2\delta_5)^3 &= \delta^3 + 3\delta^5 + 3.6\delta^7 + \ldots \\
(0.2\delta_5)^4 &= \delta^4 + 4\delta^6 + 6.8\delta^8 + \ldots \\
(0.2\delta_5)^5 &= \delta^5 + 5\delta^7 + \ldots \\
(0.2\delta_5)^6 &= \delta^6 + 6\delta^8 + \ldots \\
(0.2\delta_5)^7 &= \delta^7 + \ldots \\
(0.2\delta_5)^8 &= \delta^8 + \ldots
\end{aligned}\right\}\quad(8)$$

These expansions are, as has already been said, *finite;* and it is possible, without much trouble, to find limits to the error committed in leaving out differences above a certain order; for, an upper limit to $|f^{(\nu)}(\xi)|$ is, at the same time, an upper limit to $|\delta^\nu|$; compare the last numerical example below.

82. The application of (4) and (5) is best explained by means of a few numerical examples. In the table below, the values of a polynomial $f(x)$ of the fourth degree have been given for the arguments 50, 55, 60, 65, 70 and 75. It is desired to find the values of $f(x)$ for $x = 61$, 62, 63 and 64. It is, therefore, a question of subdividing the interval from 60 to 65 into 5 parts. For this purpose we apply (8) in the following manner:

x	$f(x)$	$0.2\delta_5$	$(0.2\delta_5)^2$	$(0.2\delta_5)^3$	$(0.2\delta_5)^4$
50	500				
		116.0			
55	1080		15.92		
		195.6		1.200	
60	2058		21.92		0.0384
		305.2		1.392	
65	3584		28.88		0.0384
		449.6		1.584	
70	5832		36.80		
		633.6			
75	9000				

x	$f(x)$	δ	δ^2	δ^3	δ^4
60	2058		21.8432		0.0384
		256.1376		1.3152	
61	2314.1376		23.1584		0.0384
		279.2960		1.3536	
62	2593.4336		24.5120		0.0384
		303.8080		1.3920	
63	2897.2416		25.9040		0.0384
		329.7120		1.4304	
64	3226.9536		27.3344		0.0384
		357.0464		1.4688	
65	3584		28.8032		0.0384

We begin by calculating the quantities $0.2\delta_5$, $(0.2\delta_5)^2$, etc.; each of these is calculated from the preceding column by subtracting two neighbouring values from each other and multiplying the result by 0.2 (with a little practice these two operations may be performed simultaneously).

Of the table thus obtained, use is only made of the underlined figures. By means of these, we form the underlined figures in the next table, beginning from the right, the working equations being, by (8),

$$\delta^4 = (0.2\delta_5)^4,\ \delta^3 = (0.2\delta_5)^3,\ \delta^2 = (0.2\delta_5)^2 - 2\delta^4,\ \delta = 0.2\delta_5 - \delta^3,$$

as all differences of higher order than the fourth vanish, our polynomial being of the fourth degree.

As δ^4 is a constant, the column headed δ^4 may at once be filled up. Next, we fill up the column δ^3, from the known middle value upwards by subtraction, and downwards by addition, of δ^4. The column δ^2 is filled up, as regards the upper half, by addition, and, as regards the lower half, by subtraction, of δ^3; as a check we have that the difference between the two middle values must be 1.3920. In this way we may continue, until the column headed $f(x)$ has been filled up.

83. At the beginning and end of a table this method cannot always be employed, as central differences are not usually available here. In such cases it becomes necessary to interpolate by the usual methods for each separate value of the function.

84. The subdivision into an *even* number of parts is usually a little more complicated, as in this case $\square\ \delta^\nu$ occurs in the formulas. As an example we give a table of a polynomial of the fourth degree and propose to halve the intervals from 0.30 to 0.80.

x	$f(x)$	$\frac{1}{2}\delta_2$	$(\frac{1}{2}\delta_2)^2$	$(\frac{1}{2}\delta_2)^3$	$(\frac{1}{2}\delta_2)^4$
0.10	2.9401				
		−0.08925			
0.20	2.7616		−0.02875		
		−0.14675		0.00075	
0.30	2.4681		−0.02725		0.00015
		−0.20125		0.00105	
0.40	2.0656		−0.02515		15
		−0.25155		0.00135	
0.50	1.5625		−0.02245		15
		−0.29645		0.00165	
0.60	0.9696		−0.01915		15
		−0.33475		0.00195	
0.70	0.3001		−0.01525		15
		−0.36525		0.00225	
0.80	−0.4304		−0.01075		15
		−0.38675		0.00255	
0.90	−1.2039		−0.00565		
		−0.39805			
1.00	−2.0000				

x	$f(x)$	δ	$\square\delta$	δ^2	δ^3	$\square\delta^3$	δ^4
0.30	2.4681			−0.0272875			0.00015
		−0.18809375			0.000975		
0.35	2.28000625		−0.20125	−0.0263125		0.00105	15
		−0.21440625			0.001125		
0.40	2.0656			−0.0251875			15
		−0.23959375			0.001275		
0.45	1.82600625		−0.25155	−0.0239125		0.00135	15
		−0.26350625			0.001425		
0.50	1.5625			−0.0224875			15
		−0.28599375			0.001575		
0.55	1.27650625		−0.29645	−0.0209125		0.00165	15
		−0.30690625			0.001725		
0.60	0.9696			−0.0191875			15
		−0.32609375			0.001875		
0.65	0.64350625		−0.33475	−0.0173125		0.00195	15
		−0.34340625			0.002025		
0.70	0.3001			−0.0152875			15
		−0.35869375			0.002175		
0.75	−0.05859375		−0.36525	−0.0131125		0.00225	15
		−0.37180625			0.002325		
0.80	−0.4304			−0.0107875			15

We begin by forming the columns headed $(\frac{1}{2}\,\delta_2)^\nu$, as shown in the table. The underlined figures in the following table are obtained (from right to left) by the formulas

$$\delta^4 = (\tfrac{1}{2}\,\delta_2)^4,\ \square\,\delta^3 = (\tfrac{1}{2}\,\delta_2)^3,\ \delta^2 = (\tfrac{1}{2}\,\delta_2)^2 - \tfrac{1}{4}\,\delta^4,\ \square\,\delta = \tfrac{1}{2}\,\delta_2,$$

resulting from (6). As δ^4 is constant, this column may immediately be filled up. We may, next, obtain δ^3 by means of $\square\,\delta^3$ and the formula

$$E^{\pm\frac{1}{2}}\delta^r = \square\,\delta^r \pm \tfrac{1}{2}\,\delta^{r+1}, \tag{9}$$

the numbers in the column of δ^3 being displaced half an interval in comparison with $\square\,\delta^3$. Thereafter, we fill up the column δ^2 by means of δ^3, the previously calculated values of δ^2 serving as a check on the calculation. We are, now, able to form δ by means of $\square\,\delta$ and δ^2, making use of (9); finally the required values of $f(x)$ are found by δ.

In practice, δ^ν and $\square\,\delta^\nu$ are written in the same column, in order to save space; here, we have placed them in separate columns in order to make the explanation easier.

85. There is another way of halving intervals which is often preferable, especially when an arithmometer is available. If, in Bessel's formula, we put $x = 0$, we obtain, leaving out the remainder-term, the symbolical formula

$$\left.\begin{aligned} 1 &= \Box - \frac{1}{2!\,2^2}\,\Box\,\delta^2 + \frac{3^2}{4!\,2^4}\,\Box\,\delta^4 - \frac{3^2 5^2}{6!\,2^6}\,\Box\,\delta^6 + \cdots \\ &= \Box\left(1 - \frac{1}{8}\,\delta^2 + \frac{3}{128}\,\delta^4 - \frac{5}{1024}\,\delta^6 + \cdots\right) \end{aligned}\right\} \quad (10)$$

The application of this important formula to the case in hand can be seen from the table below and leads, of course, to the same result as before.

x	$f(x)$	δ	δ^2	δ^3	δ^4	$1 - \frac{\delta^2}{8} + \frac{3\delta^4}{128}$	$\Box\left(1 - \frac{\delta^2}{8} + \frac{3\delta^4}{128}\right)$
0.10	2.9401						
		−0.1785					
0.20	2.7616		−0.1150				
		−0.2935		0.0060			
0.30	2.4681		−0.1090		0.0024	2.48178125	
		−0.4025		0.0084			2.28000625
0.40	2.0656		−0.1006		24	2.07823125	
		−0.5031		0.0108			1.82600625
0.50	1.5625		−0.0898		24	1.57378125	
		−0.5929		0.0132			1.27650625
0.60	0.9696		−0.0766		24	0.97923125	
		−0.6695		0.0156			0.64350625
0.70	0.3001		−0.0610		24	0.30778125	
		−0.7305		0.0180			−0.05859375
0.80	−0.4304		−0.0430		24	−0.42496875	
		−0.7735		0.0204			
0.90	−1.2039		−0.0226				
		−0.7961					
1.00	−2.0000						

86. The kind of subdivision of intervals, most frequently occurring, is the subdivision of intervals into ten equal parts. It may, of course, be done directly, but it is more convenient first to perform a halving, and then a subdivision into five parts, of the interval. Generally speaking, if a subdivision into an even number of intervals is to be performed, it is preferable to begin with the required number of halvings and, thereafter, subdivide into an odd number of intervals.

87. If the given values of the function are not exact, but approximate, numbers, the method remains, in principle, the same. The question of interpolating under these circumstances was dealt with in §5, and we may, on the whole, refer to the observations made there. As a guide to the treatment of the remainder-term in the formulas (4) and (5), if we stop at differences of a certain order, we shall, however, add one more example.

x	$\log x$	$\frac{1}{3}\delta_3$	$(\frac{1}{3}\delta_3)^2$	$(\frac{1}{3}\delta_3)^3$	$(\frac{1}{3}\delta_3)^4$
50	1.698970				
		0.0084353			
53	1.724276		−1549		
		79707		54	
56	1.748188		−1387		−3
		75547		46	
59	1.770852		−1249		−2
		71800		39	
62	1.792392		−1132		
		68403			
65	1.812913				

x	$\log x$	δ	δ^2	δ^3	δ^4
56	1.748188		−1385		−3
		0.0076868		49	
57	1.755875		−1336		−3
		0.0075532		46	
58	1.763428		−1291		−2
		0.0074241		44	
59	1.770852		−1247		−2

In the above table, the values of $\log x$ with six decimals have been given for $x = 50, 53, 56, 59, 62$ and 65; it is desired to subdivide the interval from 56 to 59 into three equal parts.

From $f(x) = \log x$ we find

$$f^{(\nu)}(x) = (-1)^{\nu-1}\frac{M(\nu-1)!}{x^\nu} \qquad (M = 0.43429448\ldots)$$

so that

$$|\delta^\nu| = |f^{(\nu)}(\xi)| \leq \frac{M(\nu-1)!}{50^\nu} < \frac{2^{\nu-1}(\nu-1)!}{10^{2\nu}}.$$

If, in (7), we neglect fifth and higher differences, this system of equations assumes the form

$$\left.\begin{aligned} \tfrac{1}{3}\,\delta_3 &= \delta + \tfrac{1}{3}\,\delta^3 \\ (\tfrac{1}{3}\,\delta_3)^2 &= \delta^2 + \tfrac{2}{3}\,\delta^4 \\ (\tfrac{1}{3}\,\delta_3)^3 &= \delta^3 \\ (\tfrac{1}{3}\,\delta_3)^4 &= \delta^4. \end{aligned}\right\} \quad (11)$$

The first of these equations is exact. In the following ones, the errors are respectively

$$\left.\begin{aligned} &\tfrac{1}{9}\,\delta^6 \\ &\delta^5 + \tfrac{1}{3}\,\delta^7 + \tfrac{1}{27}\,\delta^9 \\ &\tfrac{4}{3}\,\delta^6 + \tfrac{2}{3}\,\delta^8 + \tfrac{4}{27}\,\delta^{10} + \tfrac{1}{81}\,\delta^{12}. \end{aligned}\right\} \quad (12)$$

Now, $|\delta^5| < \dfrac{384}{10^{10}}$, or less than 4 units of the 8th decimal, and the upper limit to the following differences as far as δ^{12} is still smaller. The error committed in using (11) for the calculation of δ^ν can, therefore, as appears from the coefficients in (12), not influence the sixth decimal and is, consequently, within the errors already present in the given values of the function. The calculation is, therefore, performed exactly as if all the figures were exact; we refer for details to the table above, where the resulting values of log 57 and log 58 are only influenced by the errors present in the last decimal of the given values of log x.

We need hardly add that the methods given in this section are chiefly of importance in the case of the construction of tables of comparatively great extent.

§ 9. Inverse Interpolation

88. The problem we have solved in §4 is, to calculate the value of $f(x)$ corresponding to a given x. The opposite problem: to calculate the argument, the value of the function being given, is called *inverse interpolation.* In that case, the given value of the function is usually zero; whether this be so or not, it is seen that inverse interpolation is equivalent to the determination of a root of an equation. This is, in fact, the practical point of view which, therefore, we will adopt. We do not, however, propose to occupy ourselves with all the methods which have been invented for the numerical calculation of the roots of an equation.[1] We confine ourselves to the methods founded on the principles of the theory of interpolation from which follows, for instance, that we shall only occupy ourselves with the calculation of the *real* roots of the equation.

89. Theoretically, the problem may be said to have been solved already by *Newton's formula with divided differences.* For, if we have to do with divided differences, argument and function are only two names for the same thing, and the general formula remains the same, if they are exchanged. We need, therefore, only take $f(x)$ as argument, x as function, and form the divided differences, after which the problem is one of direct interpolation. But in practice there is the difference that even if $f(x)$ is a simple function, the inverse function may be troublesome, e.g. be many-valued, possess discontinuities, etc., so that the remainder-term causes difficulties. We therefore formulate the problem in the following way.

Let the given equation be $\varphi(y) = 0$. We put $x = \varphi(y)$ whence $y = f(x)$, so that f is the inverse function to φ. We assume that we possess a number of values of x corresponding to given values of y; these we put together in a table of $f(x)$. We now commence by examining whether $f(x)$, within an interval containing $x = 0$ and the given values of x, is real, single-valued and possesses continuous differential coefficients up to and including a certain order $n + 1$. If this is the case, we may apply Newton's formula and find for the required root, putting $x = 0$, in accordance with §4 (8)

$$\left.\begin{aligned} f(0) &= f(a_0) - a_0 f_1 + a_0 a_1 f_2 - \ldots + (-1)^n a_0 a_1 \ldots a_{n-1} f_n + R \\ R &= (-1)^{n+1} a_0 a_1 \ldots a_n \frac{f^{(n+1)}(\xi)}{(n+1)!}. \end{aligned}\right\} \quad (1)$$

90. If φ, as is often the case, is a comparatively simple function, we may, instead of applying the remainder-term, insert the value of the root found in the given equation and, in this way, satisfy ourselves that the approximation obtained is sufficient.

[1] See, for instance, Whittaker and Robinson: The Calculus of Observations, Chapter VI, or: Runge und König: Numerisches Rechnen, §§49–56.

91. As an example, let us assume that the value of an annuity-certain for 30 years is 20, the annual payment being 1, and let us attempt to calculate the rate of interest implied. The equation to be solved by (1) is, then,

$$\varphi(y) \equiv \frac{1-(1+y)^{-30}}{y} - 20 = 0.$$

Now we find in an interest-table

Rate of interest	30-years annuity
$2\frac{1}{2}\%$	20.9303
3%	19.6004
$3\frac{1}{2}\%$	18.3920

If, therefore, $\varphi(y)$ is denoted by x, and y by $f(x)$, we may form the following table

x	$f(x)$	f_1	f_2
0.9303	0.025		
		−0.0037597	
−0.3996	0.030		0.00014892
		−0.0041377	
−1.6080	0.035		

It is not difficult to ascertain that, within the interval considered, $f(x)$ satisfies the conditions mentioned above, so that (1) may be applied. We thus obtain, leaving aside the remainder-term,

$$f(0) = 0.025 + 0.9303 \times 0.0037597 - 0.9303 \times 0.3996 \times 0.00014892$$
$$= 0.028442,$$

being the required rate of interest.

If we insert this value in the equation, we see that it is a trifle too small, while it is found that 0.028450 is a trifle too large. The correct value to 6 places of decimals is, in fact, 0.028446.

92. In cases where the interval is so small that we need not go further than the second difference, the problem is sometimes, in order to avoid the divided differences, treated by *forming the ordi-*

nary difference-table and solving an equation of the second degree. Using the same example, we form the table

y	$\varphi(y)$	Δ	Δ^2
0.025	0.9303		
		−1.3299	
0.030	−0.3996		0.1215
		−1.2084	
0.035	−1.6080		

Putting $y = 0.025 + 0.005x$, we find by the interpolation-formula with descending differences, leaving aside the remainder-term,

$$0 = 0.9303 - 1.3299x + 0.1215 \frac{x(x-1)}{2}$$

or

$$0.06075x^2 - 1.39065x + 0.9303 = 0,$$

whence

$$x = 0.6898$$
$$y = 0.025 + 0.6898 \times 0.005$$
$$= 0.028449,$$

being slightly too large. Here, as before, it is necessary, if the remainder-term is left aside, to test, by insertion in the given equation, how many figures of the result can be relied upon.

93. The most frequently applied method is, however, neither of these, but simply *repeated application of linear interpolation.* In this case, too, the remainder-term is left aside, and the degree of approximation obtained is controlled by insertion of the result in the given equation.

As an example, let us calculate the positive root of the equation

$$e^x - \frac{1}{x} = 0.$$

There is evidently only one such root. In order to find it, we may leave the equation as it stands; but it is slightly more convenient to write the equation in the form

$$x \log e + \log x = 0.$$

We now put

$$y = x \log e + \log x$$

and calculate, step by step, the following table

x	y
0.5	−0.0839
0.6	0.0387
0.5684	0.0015
0.5671	−0.000052
0.5672	0.000068

It is first found, by trial, that the root is situated between 0.5 and 0.6. In thus locating the root, we compute with the smallest possible number of figures; two figures will do. The two first values of y are now calculated to four places. Thereafter, the next argument x is calculated by linear interpolation, that is

$$x = 0.5 + 0.1 \frac{0.0839}{0.1226} = 0.5684,$$

and the corresponding value of y can be calculated to four places.

By means of the last value of x found and the one that is nearest to it, or $x = 0.6$, we interpolate a new x

$$x = 0.5684 - 0.0015 \frac{0.0316}{0.0372} = 0.5671,$$

and the corresponding value of y is calculated to six places.

It is seen that we are now very close to the desired root, and that this must be either 0.5671 or 0.5672, if we are content with four figures in the root. We therefore need not interpolate any more, but in order to decide between the two values, we still calculate the value of y corresponding to $x = 0.5672$, and find that $x = 0.5671$ is preferable.

In this way, the approximation may evidently be carried as far, as we wish. At the same time we have, in the direct calculation of y, a most efficient check on the correctness of the calculation. It is evidently of importance, not to make the calculation with more figures than necessary and, therefore, to begin with few figures and only to introduce more figures when they are actually required.

94. A fourth method makes use of *divided differences with repeated argument.*[1] This method is, however, only applicable in the case of algebraical equations and, practically, only if these are of a low degree, while the methods given above are also applicable to transcendental equations.

We have shown (§3 (14)) that if all the arguments coincide, then

$$f_r = \frac{f^{(r)}(x_0)}{r\,!} \qquad (x_0 = x_1 = \ldots = x_r). \tag{2}$$

If, therefore, we write the equation in the form

$$f(x) \equiv f_0 + xf_1 + x^2 f_2 + \ldots + x^n f_n = 0, \tag{3}$$

then the coefficients are divided differences with the repeated argument zero. These are, then, immediately read off and inserted in the difference-table.

Let us, as an example, consider the equation

$$x^3 - 2x - 5 = 0, \tag{4}$$

an equation which is seen to possess only one real root. We may, then, immediately write down that portion of the difference-table below which corresponds to the four first identical arguments.[2]

x	$f(x)$	f_1	f_2	f_3
0	−5			
		−2		
0	−5		0	
		−2		1
0	−5		0	
		−2		1
0	−5		2	
		2		1
2	−1		4.1	
		10.61		1
2.1	0.061		6.2	
		11.23		1
2.1	0.061		6.3	
		11.23		1
2.1	0.061		6.29455	
		11.19569		
2.09455	−0.0000165			

[1] Thiele: Interpolationsrechnung §9.

[2] In practice, only the ascending line of divided differences −5, −2, 0, 1 is written.

As the third divided difference is constant, the column headed f_3 may immediately be filled up. The required root is evidently situated in the neighbourhood of 2; we therefore add the argument 2 in the argument-column. In order to calculate the corresponding part of the difference-table we note that, according to §3 (1), we have generally

$$f(x_1, x_2, \ldots x_r) = f(x_0, x_1, \ldots x_{r-1}) + (x_r - x_0)f(x_0, x_1 \ldots x_r), \quad (5)$$

a relation which is also valid for coinciding arguments.

We find, thus, in succession, f_3 being known, $f_2 = 2$, $f_1 = 2$, $f(2) = -1$, and insert these values in their places in the difference table.

We now try the argument $x = 2.1$. We find in succession, by (5), $f_2 = 4.1$, $f_1 = 10.61$, $f(2.1) = 0.061$, so that we have now to a considerable degree approached the root. We might continue in the same way, choosing the next time an argument, situated between 2 and 2.1 but closer to the latter value. But, having already got so close to the root, the calculation may be abbreviated by *repeating* the last argument, as shown in the table, so that it occurs 3 times altogether. At the first repetition, we calculate $f_2 = 6.2$, then $f_1 = 11.23$, while $f(2.1)$ is given and $= 0.061$. At the second repetition, we need only calculate $f_2 = 6.3$, f_1 and $f(2.1)$ being given. The divided differences formed with the repeated argument 2.1 are the coefficients in our equation, written in the form

$$0.061 + 11.23\,(x - 2.1) + 6.3\,(x - 2.1)^2 + (x - 2.1)^3 = 0; \quad (6)$$

if, in this, we neglect the third power of $(x - 2.1)$ and solve the quadratic, we get $x = 2.09455$.

As a check, we insert this value as the last argument and calculate, by (5), $f_2 = 6.29455$, $f_1 = 11.19569$, $f(2.09455) = -0.0000165$, showing that the value found is slightly too low. If greater accuracy is required, the argument 2.09455 may be repeated. The five decimals prove to be all correct.

95. The same method may evidently be used for the extraction of roots. In the case of extraction of square roots, towards the end of the calculation, instead of solving as above a quadratic equation, we repeat the argument once and solve a linear equation.

If, for instance, we want to extract the square root of 4819, we put

$$f(x) \equiv 4819 - x^2 = 0,$$

and the calculation proceeds as shown in the table.

x	$f(x)$	f_1	f_2
0	4819		
		0	
0	4819		−1
		0	
0	4819		−1
		−70	
70	−81		−1
		−139	
69	58		−1
		−138.4	
69.4	2.64		−1
		−138.8	
69.4	2.64		−1
		−138.819	
69.4190	0.0024		

The repetition of the argument 69.4 leads to the equation

$$2.64 - 138.8\,(x - 69.4) - (x - 69.4)^2 = 0$$

where the last term may be neglected. From the linear equation we then get $x = 69.4190$, and the calculation of the value of $f(x)$ corresponding to this argument shows that we need not go any further, unless we want more decimals in the result.

§10. Elementary Methods of Summation

96. We shall here occupy ourselves with some elementary summation problems which are solved by means of the methods of the theory of interpolation.

The problem of finding the sum

$$\sum_{\alpha}^{\beta} f(x) = f(\alpha) + f(\alpha + 1) + \ldots + f(\beta) \tag{1}$$

is evidently solved, if we know a function $\Phi(x)$, such that $\Delta\,\Phi(x) = f(x)$. For we obtain by summation of the latter equation

$$\sum_{\alpha}^{\beta} f(x) = \Phi(\beta + 1) - \Phi(\alpha). \tag{2}$$

The function $\Phi(x)$ need strictly only be defined at the points $x = \alpha, \alpha + 1, \ldots, \beta + 1$; in the intervals between these points we may leave $\Phi(x)$ undefined or define it in any manner we please; but as a rule a natural choice of a continuous function presents itself.

97. By means of these simple remarks we are enabled to solve a number of important elementary problems. Thus, from $\triangle x^{(r+1)} = (r + 1) x^{(r)}$ and the analogous relations we find immediately

$$\sum_{0}^{n-1} x^{(r)} = \sum_{r}^{n-1} x^{(r)} = \frac{n^{(r+1)}}{r+1}; \sum_{1}^{n} x^{(-r)} = \frac{n^{(-r-1)}}{r+1}; \sum_{\frac{1}{2}}^{n-\frac{1}{2}} x^{[r]} = \frac{n^{[r+1]}}{r+1}. \quad (3)$$

These relations are quite analogous to the integration-formula

$$\int_0^n x^r \, dx = \frac{n^{r+1}}{r+1}.$$

98. By means of (3) every polynomial in x may be summed. For we may, as shown in No. 14, expand the polynomial in one of the factorials $x^{(\nu)}$, $x^{(-\nu)}$ or $x^{[\nu]}$ and, thereafter, employ (3). The result may, if desired, again be expressed in ordinary powers of x.[1]

If, for instance,

$$Q(x) = \sum_{r=0}^{m} c_r x^r = \sum_{\nu=0}^{m} \frac{x^{(\nu)}}{\nu!} \triangle^{\nu} Q(0),$$

we find

$$\sum_{x=0}^{n-1} Q(x) = \sum_{\nu=0}^{m} \frac{n^{(\nu+1)}}{(\nu+1)!} \triangle^{\nu} Q(0).$$

In particular

$$\sum_{x=0}^{n-1} x^r = \sum_{\nu=0}^{r} \frac{n^{(\nu+1)}}{(\nu+1)!} \triangle^{\nu} 0^r. \quad (4)$$

More general limits of summation are easily introduced. Thus

$$\sum_{\alpha}^{\beta} Q(x) = \sum_{0}^{m} \frac{(\beta+1)^{(\nu+1)} - \alpha^{(\nu+1)}}{(\nu+1)!} \triangle^{\nu} Q(0).$$

[1] Another method of summation will be mentioned in §13.

It may be noted that the formula employed in No. 40 for the sum of the squares of the odd numbers is a particular case of the last formula (3) from which we obtain, for $r = 2$,

$$\sum_{\frac{1}{2}}^{n-\frac{1}{2}} x^{[2]} = \frac{n^{[3]}}{3}$$

or

$$1^2 + 3^2 + \dots + (2n-1)^2 = \frac{n(4n^2-1)}{3}. \tag{5}$$

This formula may, of course, also be derived by putting

$$Q(x) = (2x+1)^2 = 1 + 4x + 4x^2$$

and using any of the above methods.

99. The same method of summation may be used with respect to the *inverse factorial*, or

$$\frac{1}{x(x+1)\dots(x+r-1)}.$$

For we have

$$\Delta \frac{1}{x(x+1)\dots(x+r-1)} = \frac{-r}{x(x+1)\dots(x+r)}, \tag{6}$$

and hence

$$\sum_{n}^{\infty} \frac{1}{x(x+1)\dots(x+r)} = \frac{1}{r} \cdot \frac{1}{n(n+1)\dots(n+r-1)}. \tag{7}$$

From this, an expression for the sum, taken between arbitrary limits, is easily derived; we need not write it down.

100. Expressions of the form

$$\frac{Q(x)}{x(x+1)\dots(x+r)} \tag{8}$$

may sometimes be summed, as is seen by expanding $Q(x)$ in ascending factorials. If, however, this leads to terms of the form $\frac{a}{x+b}$, the summation can only be performed with approximation, by means of methods to be developed later on. Corresponding remarks may be made concerning expressions of the form

$$\frac{Q(x)}{(x+n_1)(x+n_2)\ldots(x+n_r)}, \tag{9}$$

the numbers n_ν being integers. If the numerator and denominator are multiplied by suitable factors, we obtain the form (8).

101. Many trigonometrical expressions may be summed by similar means. Thus, by summing the identity

$$\Delta \sin (x-\tfrac{1}{2})\, a = 2 \sin \frac{a}{2} \cos xa$$

we obtain the formula, used in the theory of the trigonometrical series,

$$\frac{1}{2} + \sum_{1}^{n} \cos xa = \frac{\sin (n+\frac{1}{2})\, a}{2 \sin \dfrac{a}{2}}. \tag{10}$$

102. A general formula which is of great use in many summation problems, is the formula for *partial summation*, which is analogous to the formula for partial integration. We obtain immediately, from the identity

$$\Delta u_x v_x = u_x \Delta v_x + v_{x+1} \Delta u_x, \tag{11}$$

the desired formula

$$\sum_{0}^{n-1} u_x \Delta v_x = \Big[u_x v_x \Big]_0^n - \sum_{0}^{n-1} v_{x+1} \Delta u_x. \tag{12}$$

103. As an application of this formula we will show how expressions of the form $a^x Q(x)$ can be summed, $Q(x)$ denoting, as before, a polynomial.

We have, for $a \neq 1$,

$$\sum_{0}^{n-1} a^x Q(x) = \frac{1}{a-1} \sum_{0}^{n-1} Q(x) \Delta a^x,$$

hence, if, in (12), we put $u_x = Q(x)$, $v_x = a^x$,

$$\sum_{0}^{n-1} a^x Q(x) = \frac{1}{a-1}\left\{\left[a^x Q(x)\right]_0^n - \sum_{0}^{n-1} a^{x+1}\Delta Q(x)\right\}. \tag{13}$$

By means of this formula, the problem is reduced to the summation of $a^x \Delta Q(x)$, the polynomial by which a^x is multiplied being one degree lower than in the original expression. The new summation-problem may, in the same way, be reduced to the summation of $a^x \Delta^2 Q(x)$, and so on, until we arrive at the summation of a^x which is immediately performed.

If, by $\Sigma f(x)$ or the *indefinite sum* (an analogue of the indefinite integral), we understand the sum from a constant but not specified number to $x - 1$, we have, according to (13),

$$\Sigma\, a^x Q(x) = \frac{a^x}{a-1} Q(x) - \frac{a}{a-1} \Sigma\, a^x \Delta Q(x) + C, \tag{14}$$

C being a constant which, as in the integral calculus, is determined by the insertion of limits. We then find, by repeated application of (14), if $Q(x)$ has the degree r,

$$\Sigma\, a^x Q(x) = \frac{a^x}{a-1} \sum_{0}^{r} (-1)^\nu \left(\frac{a}{a-1}\right)^\nu \Delta^\nu Q(x) + C. \tag{15}$$

If, for instance, $Q(x) = x^{(r)}$, we get, from (15),

$$\Sigma\, a^x x^{(r)} = \frac{a^x}{a-1} \sum_{\nu=0}^{r} (-1)^\nu \left(\frac{a}{a-1}\right)^\nu r^{(\nu)} x^{(r-\nu)} + C,$$

whence

$$\sum_{0}^{n-1} a^x x^{(r)} = \frac{a^n}{a-1} \sum_{0}^{r} (-1)^\nu \left(\frac{a}{a-1}\right)^\nu r^{(\nu)} n^{(r-\nu)} + (-1)^{r+1} \left(\frac{a}{a-1}\right)^r \frac{r!}{a-1}. \tag{16}$$

We may, of course, also use this formula for summing any polynomial, multiplied by a^x, if we begin by expanding the polynomial in descending factorials.

104. Functions of the form

$$a^x \cos bx.Q(x) \quad \text{or} \quad a^x \sin bx.Q(x) \tag{17}$$

may be summed by a similar process; most simply by expressing $\cos bx$ and $\sin bx$ by means of exponentials.

105. Most of the formulas derived here may be generalized by a linear transformation of the argument, putting $x = h + ky$. We shall not go into details, but confine ourselves to pointing out, that if $u_x = ax + b$, then

$$\Delta u_x u_{x+1} \dots u_{x+r} = a(r+1) u_{x+1} u_{x+2} \dots u_{x+r} \tag{18}$$

and consequently

$$\Sigma u_{x+1} u_{x+2} \dots u_{x+r} = \frac{1}{a(r+1)} u_x u_{x+1} \dots u_{x+r} + C; \tag{19}$$

further

$$\Delta \frac{1}{u_x u_{x+1} \dots u_{x+r}} = \frac{-a(r+1)}{u_x u_{x+1} \dots u_{x+r+1}}, \tag{20}$$

and hence

$$\Sigma \frac{1}{u_x u_{x+1} \dots u_{x+r+1}} = -\frac{1}{a(r+1)} \cdot \frac{1}{u_x u_{x+1} \dots u_{x+r}} + C, \tag{21}$$

(19) and (21) corresponding to the second equation (3) and (7) respectively.

106. A theorem which is often of great use for the transformation of summation-problems is the identity

$$\sum_{y=\alpha}^{\beta} \sum_{x=\alpha}^{y} w_{x,y} = \sum_{x=\alpha}^{\beta} \sum_{y=x}^{\beta} w_{x,y} \tag{22}$$

which is called Dirichlet's sum-formula.

In order to prove it, we note that the left-hand side, written in full, is

$$\begin{array}{l} w_{\alpha,\alpha} \\ + w_{\alpha,\alpha+1} + w_{\alpha+1,\alpha+1} \\ + w_{\alpha,\alpha+2} + w_{\alpha+1,\alpha+2} + w_{\alpha+2,\alpha+2} \\ + \dots\dots\dots\dots\dots\dots \\ + w_{\alpha,\beta} \quad + w_{\alpha+1,\beta} \quad + w_{\alpha+2,\beta} + \dots + w_{\beta,\beta}, \end{array}$$

while the right-hand side is

$$\begin{array}{llll} w_{\alpha,\alpha} + w_{\alpha,\alpha+1} & + w_{\alpha,\alpha+2} & + \cdots\cdots & + w_{\alpha,\beta} \\ \quad + w_{\alpha+1,\alpha+1} & + w_{\alpha+1,\alpha+2} & + \cdots\cdots & + w_{\alpha+1,\beta} \\ & + w_{\alpha+2,\alpha+2} & + \cdots\cdots & + w_{\alpha+2,\beta} \\ & & + \cdots\cdots\cdots\cdots & \\ & & & + w_{\beta,\beta}. \end{array}$$

But comparing columns of the latter expression with rows of the former, we see that they are identical.

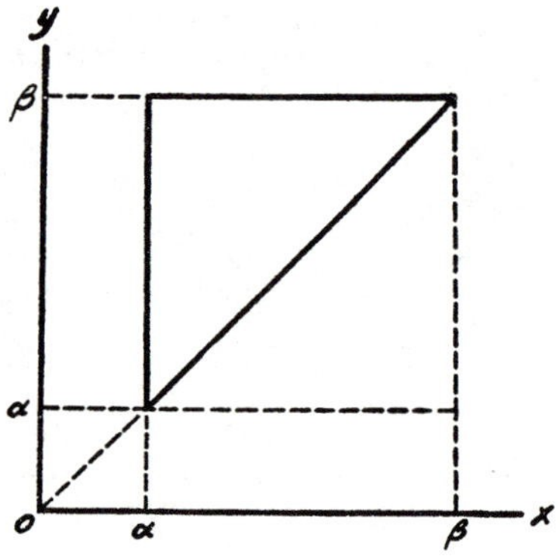

FIG. 2

A particular case of (22) is so frequently applied that it should be noted separately. It is obtained on putting $w_{x,y} = u_x v_y$, the result being

$$\sum_{\alpha}^{\beta} v_x \sum_{\alpha}^{x} u_x = \sum_{\alpha}^{\beta} u_x \sum_{x}^{\beta} v_x. \tag{23}$$

This formula is useful, when one of the functions u_x and v_x can be summed, but the other not. For instance

$$\sum_{1}^{n} x^{(-r)} \sum_{x}^{n} \frac{1}{x} = \sum_{1}^{n} \frac{1}{x} \sum_{1}^{x} x^{(-r)} = \sum_{1}^{n} \frac{1}{x} \frac{x^{(-r-1)}}{r+1}$$

$$= \frac{1}{r+1} \sum_{2}^{n+1} x^{(-r)} = \frac{(n+1)^{(-r-1)} - (r+1)!}{(r+1)^2}$$

A similar relation as (22) holds for integrals, viz. the formula

$$\int_\alpha^\beta dy \int_\alpha^y F(x;y)\,dx = \int_\alpha^\beta dx \int_x^\beta F(x;y)dy, \tag{24}$$

or Dirichlet's integral-formula. In order to prove it, we note that both sides of this equation represent the result of integrating the function $F(x; y)$ over the area consisting of a rectangular triangle, shown in figure 2, the only difference being that on the left the triangle is divided into small strips parallel to the x-axis, while on the right the strips are parallel to the y-axis.

If, in (24), we put $F(x; y) = f(x)\,\varphi(y)$, we get

$$\int_\alpha^\beta \varphi(x) \int_\alpha^x f(x)dx^2 = \int_\alpha^\beta f(x) \int_x^\beta \varphi(x)dx^2. \tag{25}$$

Dirichlet's integral-formula may also be derived from the sum-formula by a limiting process.[1]

§11. Repeated Summation

107. If there is given a table of $f(x)$

$$f(\alpha), f(\alpha+1), f(\alpha+2), \ldots$$

we may form the sums

$$\left.\begin{array}{l} f(\alpha) \\ f(\alpha) + f(\alpha+1) \\ f(\alpha) + f(\alpha+1) + f(\alpha+2) \\ \cdots\cdots\cdots\cdots\cdots\cdots \\ f(\alpha) + f(\alpha+1) + f(\alpha+2) + \ldots + f(x-1) \\ \cdots\cdots\cdots\cdots\cdots\cdots\cdots\cdots\cdots\cdots \end{array}\right\}$$

Let Σ, when no limits of summation are indicated, denote $\overset{x-1}{\underset{\alpha}{\Sigma}}$, that is

$$\Sigma f(x) = \overset{x-1}{\underset{\alpha}{\Sigma}} f(x). \tag{1}$$

[1] G. J. Lidstone: Journal of the Institute of Actuaries, vol. XLIV, p. 403.

The operation of summation Σ may, like the operation of differencing Δ, be *repeated*, and the repetitions are also here denoted by a symbolical exponent, e.g.

$$\Sigma^2 f(x) = \sum_{\alpha+1}^{x-1} \sum_{\alpha}^{x-1} f(x),$$

the first Σ on the right having $\alpha + 1$ as the lower limit of summation, as the table commences at $f(\alpha)$.

Thus we may continue and, therefore, have in the general case

$$\Sigma^r f(x) = \sum_{\alpha+r-1}^{x-1} \sum_{\alpha+r-2}^{x-1} \cdots \sum_{\alpha+1}^{x-1} \sum_{\alpha}^{x-1} f(x). \tag{2}$$

The calculation is best arranged in a table of the form shown below and called the *sum-table*, the analogy with the difference-table being obvious.

x	$f(x)$	Σ	Σ^2	Σ^3	Σ^4
a	$f(a)$				
$a+1$	$f(a+1)$	$\Sigma f(a+1)$			
$a+2$	$f(a+2)$	$\Sigma f(a+2)$	$\Sigma^2 f(a+2)$		
$a+3$	$f(a+3)$	$\Sigma f(a+3)$	$\Sigma^2 f(a+3)$	$\Sigma^3 f(a+3)$	
$a+4$	$f(a+4)$	$\Sigma f(a+4)$	$\Sigma^2 f(a+4)$	$\Sigma^3 f(a+4)$	$\Sigma^4 f(a+4)$

It is seen, that $f(x)$, $\Sigma f(x)$, $\Sigma^2 f(x)$, . . . $\Sigma^r f(x)$, . . . are all found on the same line, sloping upwards. All the numbers in the top-line, sloping downwards, are $f(\alpha)$, so that

$$\Sigma^r f(\alpha + r) = f(\alpha). \tag{3}$$

If this table is read from the right to the left, it is a difference-table. If differences and sums of $f(x)$ are wanted simultaneously, the differences are placed to the right, the sums to the left, of the column headed $f(x)$; so that Σ comes next to this, Σ^2 to the left of Σ, and so on. In this way the whole table becomes a difference-table, formed from the sums of highest order.

While the difference-table is perfectly determined by the function, the same does not apply to the sum-table, as the repeated sums depend on the argument α, at which the summation commences.

The operation Σ is to a certain extent the inverse operation of Δ, as $\Delta\Sigma = 1$. On the other hand, we have generally $\Sigma\Delta \neq 1$, as

$$\Sigma\Delta f(x) = f(x) - f(\alpha).$$

The symbol Σ is, therefore, not commutative with the difference-symbol.

108. The sum-table is sometimes given in a different form, putting $S = \overset{x}{\underset{\alpha}{\Sigma}}$, that is

$$Sf(x) = \overset{x}{\underset{\alpha}{\Sigma}} f(x) = \Sigma f(x+1),$$

$$S^2 f(x) = \overset{x}{\underset{\alpha}{\Sigma}}\overset{x}{\underset{\alpha}{\Sigma}} f(x) = \Sigma^2 f(x+2),$$

and generally, repeating $\overset{x}{\underset{\alpha}{\Sigma}}$ r times,

$$S^r f(x) = \overset{x}{\underset{\alpha}{\Sigma}} \dots \overset{x}{\underset{\alpha}{\Sigma}} f(x) = \Sigma^r f(x+r). \tag{4}$$

If the sums S are employed, $S^r f(x)$ is placed on the *horizontal* line starting at $f(x)$; all the numbers in the top horizontal line have the value $f(\alpha)$.

109. The rth sum may, like the rth difference, be expressed as a linear function of the values of $f(x)$ by means of which it is formed. In order to find this expression we note first that by Dirichlet's sum-formula §10 (23) we obtain the following formula where the summation refers to ν, and $r > 1$,

$$\overset{x}{\underset{\alpha}{\Sigma}} \binom{\nu - x - 1}{r-2} \overset{\nu}{\underset{\alpha}{\Sigma}} f(\nu) = \overset{x}{\underset{\alpha}{\Sigma}} f(\nu) \overset{x}{\underset{\nu}{\Sigma}} \binom{\nu - x - 1}{r-2}$$
$$= - \overset{x}{\underset{\alpha}{\Sigma}} \binom{\nu - x - 1}{r-1} f(\nu)$$

or, multiplying by $(-1)^{r-2}$ and exchanging the right- and left-hand sides,

$$\overset{x}{\underset{\nu=\alpha}{\Sigma}} \binom{x - \nu + r - 1}{r-1} f(\nu) = \overset{x}{\underset{\nu=\alpha}{\Sigma}} \binom{x - \nu + r - 2}{r-2} \overset{\nu}{\underset{\alpha}{\Sigma}} f(\nu). \tag{5}$$

In the expression on the left we may, therefore, diminish r by one unit if, at the same time, we replace $f(\nu)$ by $\overset{\nu}{\underset{\alpha}{\Sigma}} f(\nu)$.

The expression on the right of (5) may evidently be treated in the same way, the result being

$$\sum_{\nu=\alpha}^{x} \binom{x-\nu+r-2}{r-2} \overset{\nu}{\underset{\alpha}{\Sigma}} f(\nu) = \sum_{\nu=\alpha}^{x} \binom{x-\nu+r-3}{r-3} \overset{\nu}{\underset{\alpha}{\Sigma}} \overset{\nu}{\underset{\alpha}{\Sigma}} f(\nu),$$

and so on, until the right-hand side, as $\binom{x-\nu}{0} = 1$, assumes the value $S^r f(x)$, that is

$$S^r f(x) = \sum_{\nu=\alpha}^{x} \binom{x-\nu+r-1}{r-1} f(\nu), \tag{6}$$

this being the required formula.

From this follows, by (4),

$$\Sigma^r f(x) = \sum_{\nu=\alpha}^{x-r} \binom{x-\nu-1}{r-1} f(\nu). \tag{7}$$

In particular

$$S^r 1 = \binom{x-\alpha+r}{r}, \quad \Sigma^r 1 = \binom{x-\alpha}{r}. \tag{8}$$

110. Repeated summation may with advantage be employed for the calculation of sums of the form $\Sigma\,(x-a)^r f(x)$, the so-called *moments about the point a*. As the transition from moments about one point to moments about another is easily made,[1] we content ourselves with deriving the relations which have the simplest form, viz. the relations between Σ^r and the moments about $x-1$, defined by

$$M_r = \sum_{\nu=\alpha}^{x-1} (\nu - x + 1)^r f(\nu), \tag{9}$$

where we may, for $r>0$, take $x-2$ as the upper limit of summation.

[1] We need only write $\Sigma\,(x-a)^r f(x) = \Sigma\,[(x-b)+(b-a)]^r f(x)$ and develop in powers of $(x-b)$.

We now expand $(x - \nu - 1)^r$ in descending factorials

$$(x - \nu - 1)^r = \sum_{\mu=0}^{r} (x - \nu - 1)^{(\mu)} \frac{\Delta^\mu 0^r}{\mu!};$$

inserting this in (9) we obtain, after exchanging the order of summation,

$$(-1)^r M_r = \sum_{\mu=0}^{r} \Delta^\mu 0^r \sum_{\nu=\alpha}^{x-\mu-1} \binom{x-\nu-1}{\mu} f(\nu)$$

or, for $r > 0$,

$$(-1)^r M_r = \sum_{\mu=1}^{r} \Delta^\mu 0^r \Sigma^{\mu+1} f(x) \qquad (r > 0). \tag{10}$$

In particular we find from this, leaving out $f(x)$ on the right,

$$\left.\begin{aligned} -M_1 &= \Sigma^2 \\ M_2 &= \Sigma^2 + 2\Sigma^3 \\ -M_3 &= \Sigma^2 + 6\Sigma^3 + 6\Sigma^4 \\ M_4 &= \Sigma^2 + 14\Sigma^3 + 36\Sigma^4 + 24\Sigma^5. \end{aligned}\right\} \tag{11}$$

By this system of formulas, the tedious direct calculation of the moments has, in the main, been reduced to a summation-problem.

It is also possible to develop in central factorials; we content ourselves with referring the reader to G. F. Hardy: The Theory of the Construction of Tables of Mortality, pp. 124–128.

111. In the preceding formulas it has been assumed that the summation commences *at the top* of the table, which is preferable if the values of $f(x)$ have a number of figures that *increases* with x. But if the number of figures of $f(x)$ *decreases* with increasing x, summation *from the bottom* of the table is to be preferred. The formulas to be used for this purpose are, in principle, obtained by writing the given values of $f(x)$ in the reverse order. The corresponding summation-symbols are, denoting by β the last argument of the table,

$\Sigma' = \sum\limits_{x+1}^{\beta}$ and $S' = \sum\limits_{x}^{\beta}$.

We first obtain from (6), writing the terms in reverse order, and replacing x by β,

$$S'^r f(\beta) = f(\beta) + \binom{r}{r-1} f(\beta-1) + \binom{r+1}{r-1} f(\beta-2) + \ldots + \binom{\beta-\alpha+r-1}{r-1} f(\alpha)$$

and from this, for the rth sum from the bottom,

$$S^{(r)}f(\alpha) = f(\alpha) + \binom{r}{r-1} f(\alpha+1) + \binom{r+1}{r-1} f(\alpha+2) + \ldots + \binom{\beta-\alpha+r-1}{r-1} f(\beta)$$

whence, replacing α by x,

$$S^{(r)}f(x) = \sum_{\nu=x}^{\beta} \binom{\nu - x + r - 1}{r-1} f(\nu). \tag{12}$$

Now

$$S'f(x) = \sum_{x}^{\beta} f(x) = \Sigma' f(x-1),$$

$$S''f(x) = \sum_{x}^{\beta}\sum_{x}^{\beta} f(x) = \Sigma'' f(x-2),$$

and in the general case, repeating $\sum_{x}^{\beta}$ r times,

$$S^{(r)}f(x) = \sum_{x}^{\beta} \cdots \sum_{x}^{\beta} f(x) = \Sigma^{(r)}f(x-r). \tag{13}$$

We therefore obtain, by (12),

$$\Sigma^{(r)}f(x) = \sum_{\nu=x+r}^{\beta} \binom{\nu - x - 1}{r-1} f(\nu). \tag{14}$$

In particular

$$S^{(r)}\,1 = \binom{\beta - x + r}{r}, \qquad \Sigma^{(r)}\,1 = \binom{\beta - x}{r}. \tag{15}$$

A section of the sum-table for $\Sigma^{(r)}$ is shown below.

x	$f(x)$	Σ'	Σ''	Σ'''	Σ^{IV}
$\beta-4$	$f(\beta-4)$				
		$\Sigma'f(\beta-4)$			
$\beta-3$	$f(\beta-3)$		$\Sigma''f(\beta-4)$		
		$\Sigma'f(\beta-3)$		$\Sigma'''f(\beta-4)$	
$\beta-2$	$f(\beta-2)$		$\Sigma''f(\beta-3)$		$\Sigma^{IV}f(\beta-4)$
		$\Sigma'f(\beta-2)$		$\Sigma'''f(\beta-3)$	
$\beta-1$	$f(\beta-1)$		$\Sigma''f(\beta-2)$		
		$\Sigma'f(\beta-1)$			
β	$f(\beta)$				

The numbers in the line sloping upwards from the bottom have all the value $f(\beta)$, so that

$$\Sigma^{(r)} f(\beta - r) = f(\beta). \tag{16}$$

It is seen that $f(x)$, $\Sigma' f(x), \Sigma'' f(x), \ldots \Sigma^{(r)} f(x), \ldots$ are all found on the same line, sloping downwards.

We have evidently $\nabla\Sigma' = -1$; on the other hand $\Sigma'\nabla f(x) = f(\beta) - f(x)$, so that the symbol Σ' is not commutative with the difference-symbol.

112. We denote the moments about $x + 1$ by

$$M'_r = \sum_{\nu = x+1}^{\beta} (\nu - x - 1)^r f(\nu). \tag{17}$$

Now, let us in (14) commence the summation at $x + 1$, and in (7) continue the summation to $x - 1$, which does not alter the result, as the terms added all vanish. If, under these circumstances, we compare the expansion of (14) in powers of $(\nu - x - 1)$ with the expansion of (7) in powers of $(x - \nu - 1)$, we observe that they only differ from each other in that M_r has been replaced by $(-1)^r M'_r$. We therefore have, by (10),

$$M'_r = \sum_{\mu = 1}^{r} \Delta^\mu 0^r \, \Sigma^{(\mu + 1)} f(x) \quad (r > 0) \tag{18}$$

and in particular, leaving out $f(x)$,

$$\left.\begin{aligned} M'_1 &= \Sigma'' \\ M'_2 &= \Sigma'' + 2\,\Sigma''' \\ M'_3 &= \Sigma'' + 6\,\Sigma''' + 6\,\Sigma^{IV} \\ M'_4 &= \Sigma'' + 14\,\Sigma''' + 36\,\Sigma^{IV} + 24\,\Sigma^{V}. \end{aligned}\right\} \tag{19}$$

113. This system of formulas is used if the moments are to be calculated for a table of $f(x)$ with a decreasing number of figures. But very often, especially in dealing with frequency-distributions, we have to do with a table of numbers which begin by increasing, attain a maximum, and thereafter decrease. In such cases, the moments are calculated separately for the increasing and the decreasing part of the table about the same, conveniently chosen, point in the neighbourhood of the maximum. If, now, by M_r

we understand the moments for the whole table (obtained by adding the moments for the upper and lower parts of the same), and if we write, for abbreviation,

$$\left.\begin{aligned} \Sigma_r &= \Sigma^{(r)} + \Sigma^r, \\ \Delta_r &= \Sigma^{(r)} - \Sigma^r, \end{aligned}\right\} \quad (20)$$

we obtain from (19) and (11)

$$\left.\begin{aligned} M_1 &= \Delta_2 \\ M_2 &= \Sigma_2 + 2\,\Sigma_3 \\ M_3 &= \Delta_2 + 6\,\Delta_3 + 6\,\Delta_4 \\ M_4 &= \Sigma_2 + 14\,\Sigma_3 + 36\,\Sigma_4 + 24\,\Sigma_5. \end{aligned}\right\} \quad (21)$$

114. The process of calculation is best explained by a numerical example.

x	$f(x)$	1. Sum	2. Sum	3. Sum	4. Sum	5. Sum
0	15					
		15				
			15			
1	209					
		224		15		
			239		15	
2	365					
		589		254		15
			828		269	
3	482					
		1071		1082		
			1899			
4	414					
			901			
		515		592		
5	277					
			386		280	
		238		206		92
6	134					
			148		74	
		104		58		—
7	72					
			44		16	
		32		14		—
8	22					
			12		2	
		10		2		
9	8					
			2			
		2				
10	2					
$\Sigma_r =$			2800	1674	549	107
$\Delta_r =$			−998	−490	11	77

Given the above table of $f(x)$, we want to calculate the moments of this function about $x = 4$.

Above the horizontal line starting at $f(4)$, we sum from the top; below the line, from the bottom; in the last column only one number is inserted above and below the line respectively, as no more are wanted. The numbers wanted for the calculation of Σ_r and Δ_r are those placed nearest above and below the horizontal line. Thus, for instance, $\Sigma_3 = 1082 + 592 = 1674$, $\Delta_3 = 592 - 1082 = -490$, etc. The moments about $x = 4$ are now calculated by (21); we find $M_1 = -998$, $M_2 = 6148$, $M_3 = -3872$, $M_4 = 48568$.

115. It should be noted that the so-called *factorial moments* used in connection with frequency-distributions[1] and defined by

$$\bar{\sigma}_{r)} = \sum_{\nu=r}^{k} \nu^{(r)} f(\nu) = r! \sum_{\nu=r}^{k} \binom{\nu}{r} f(\nu) \tag{22}$$

are, if we leave the factor $r!$ aside, expressible as the $(r+1)^{st}$ sum from the bottom of the values of $f(0), f(1), \ldots f(k)$, that is

$$\Sigma^{(r+1)} f(-1) = \frac{\bar{\sigma}_{(r)}}{r!}. \tag{23}$$

In the sum-table for summation from the bottom, all the values $\dfrac{\bar{\sigma}_{(r)}}{r!}$ are found on the same downward sloping line starting from $f(-1)$.

116. Another field of application for repeated summation is the calculation of *product-sums*, that is, sums of the form $\Sigma v_x u_x$. If we put

$$\Sigma' u_x = u_{x+1} + u_{x+2} + \ldots + u_{n-1}, \qquad \Sigma' u_{n-1} = 0,$$

we find by partial summation

$$\sum_0^{n-1} v_x u_x = -\sum_0^{n-1} v_x \Delta \Sigma' u_{x-1}$$

$$= v_0 \Sigma' u_{-1} + \sum_0^{n-2} \Delta v_x \Sigma' u_x$$

[1] J. F. Steffensen: Factorial Moments and Discontinuous Frequency-Functions. Skandinavisk Aktuarietidskrift, 1923, p. 73.

and hence, continuing in the same way,

$$\sum_{0}^{n-1} v_x u_x = \sum_{\nu=0}^{k} \Delta^\nu v_0 \, \Sigma^{(\nu+1)} u_{-1} + R \tag{24}$$

where

$$R = \sum_{x=0}^{n-k-2} \Delta^{k+1} v_x \, \Sigma^{(k+1)} u_x. \tag{25}$$

In this formula, $\Sigma^{(\nu+1)} u_{-1}$ means the $(\nu + 1)^{st}$ sum from the bottom of the values $u_0, u_1, \ldots u_{n-1}$. These sums are all found on the same downward sloping line, starting from u_{-1}.

In order to give the remainder-term a practical form, something must be known about u_x. In practice, u_x is very often a *positive* function; in that case $\Sigma^{(k+1)} u_x$ does not change its sign, and we have, by the Theorem of Mean Value

$$R = \mu \Delta^{k+1} v_x \cdot \Sigma^{(k+2)} u_{-1} \qquad (u_x > 0), \tag{26}$$

denoting by $\mu \Delta^{k+1} v_x$ a number situated between the smallest and the largest of the differences $\Delta^{k+1} v_0, \Delta^{k+1} v_1, \ldots \Delta^{k+1} v_{n-k-2}$.

If v_x possesses a continuous derivative of order $k + 1$, the remainder-term may be written

$$R = v_\xi^{(k+1)} \, \Sigma^{(k+2)} u_{-1} \qquad (u_x > 0). \tag{27}$$

But the previous formulas possess the merit of remaining valid, even if u_x and v_x do not follow any particular law.

The application of the Error-Test leads to the result that if u_x, $\Delta^{k+1} v_x$ and $\Delta^{k+2} v_x$ do not change their signs, and if $\Delta^{k+1} v_x$ and $\Delta^{k+2} v_x$ have opposite signs, then R is numerically smaller than the first neglected term and has the same sign (provided this term does not vanish). This also holds for $n \to \infty$, if the sums are convergent.

It may be noted that for $u_x = 1$ we get the summation-formula

$$\sum_{0}^{n-1} v_x = \sum_{\nu=0}^{k} \binom{n}{\nu+1} \Delta^\nu v_0 + \binom{n}{k+2} \mu \Delta^{k+1} v_x, \tag{28}$$

the first term on the right being $n v_0$. This formula is, however, less accurate than certain summation-formulas which will be developed later on.

The subject of product-sums may be developed in several directions; for instance by introducing central differences. We shall, however, not go into this.

117. *Repeated integration* may be treated on the same lines as repeated summation. If, for instance, in Dirichlet's formula §10 (25) we put $\varphi(x) = (\beta - x)^{k-1}$, we get

$$\int_\alpha^\beta (\beta - x)^k f(x)\,dx = k\int_\alpha^\beta (\beta - x)^{k-1}\int_\alpha^x f(x)\,dx^2$$

and by repeated application of this, proceeding exactly as we did in deriving (6) from (5),

$$\int_\alpha^\beta\int_\alpha^x \dots \int_\alpha^x f(x)dx^{k+1} = \frac{1}{k!}\int_\alpha^\beta (\beta - x)^k f(x)dx. \tag{29}$$

If, on the other hand, we put, in Dirichlet's formula, $f(x) = (x - \alpha)^{k-1}$, we find

$$\int_\alpha^\beta (x - \alpha)^k \varphi(x)dx = k\int_\alpha^\beta (x - \alpha)^{k-1}\int_x^\beta \varphi(x)\,dx^2$$

and by repeated application of this

$$\int_\alpha^\beta\int_x^\beta \dots \int_x^\beta \varphi(x)dx^{k+1} = \frac{1}{k!}\int_\alpha^\beta (x - \alpha)^k \varphi(x)dx. \tag{30}$$

In this way, the calculation of repeated integrals has been reduced to a single integration.

118. It is not difficult to see how we must proceed, if the constant limits of integration in (29) and (30) are not the same in all the integrations. If, for instance, we want to calculate

$$\int_\alpha^\beta\int_\gamma^x f(x)dx^2,$$

we may note that $\int_\gamma^x = \int_\alpha^x - \int_\alpha^\gamma$, and find therefore

$$\int_\alpha^\beta\int_\gamma^x f(x)dx^2 = \int_\alpha^\beta\int_\alpha^x f(x)dx^2 - \int_\alpha^\beta\int_\alpha^\gamma f(x)dx^2$$

$$= \int_\alpha^\beta (\beta - x)f(x)dx - (\beta - \alpha)\int_\alpha^\gamma f(x)dx.$$

We may proceed in a similar way, if it is a question of sums instead of integrals.

§12. Laplace's and Gauss's Summation-Formulas

119. As an integral is the limiting value of a sum, approximate relations must exist between sums and integrals, that is, an integral is equal to a sum *plus* certain corrective terms which may be expressed in different forms. If the corrective terms are put into the form of *differences*, we have formulas of Laplace's type; if they are expressed as *differential coefficients*, we have Euler's type. Both types may be used either for the calculation of sums (if the integral is known) or for the calculation of integrals (if the sum is known).

120. Denoting by m a positive integer, we start from the identity

$$\int_0^m f(x)dx = \sum_{r=0}^{m-1} \int_0^1 f(t+r)dt. \tag{1}$$

If, in this, we express $f(t+r)$ by an expansion in descending factorials

$$f(t+r) = \sum_0^n \frac{t^{(\nu)}}{\nu!} \Delta^\nu f(r) + \frac{t^{(n+1)}}{(n+1)!} f^{(n+1)}(\xi),$$

we get

$$\int_0^m f(x)dx = \sum_{r=0}^{m-1} \int_0^1 dt \sum_{\nu=0}^{n} \frac{t^{(\nu)}}{\nu!} \Delta^\nu f(r) + R,$$

$$R = \sum_{r=0}^{m-1} \int_0^1 dt \frac{t^{(n+1)}}{(n+1)!} f^{(n+1)}(\xi).$$

If we introduce the notation

$$L_\nu = \int_0^1 \frac{t^{(\nu)}}{\nu!} dt, \tag{2}$$

we find

$$\int_0^m f(x)dx = \sum_{\nu=0}^{n} L_\nu \sum_{r=0}^{m-1} \Delta^\nu f(r) + R$$

$$= L_0 \sum_{r=0}^{m-1} f(r) + \sum_{\nu=1}^{n} L_\nu \Big[\Delta^{\nu-1} f(x) \Big]_0^m + R$$

and for the remainder-term, as $t^{(n+1)}$ does not change its sign within the interval $0 < t < 1$,

$$R = \sum_{r=0}^{m-1} f^{(n+1)}(\xi) L_{n+1} = mL_{n+1} f^{(n+1)}(\xi),$$

as, instead of $\frac{1}{m} \sum_{r=0}^{m-1} f^{(n+1)}(\xi)$ we may write $f^{(n+1)}(\xi)$ (with another ξ).

As evidently $L_0 = 1$, the result may be written in the form

$$\left.\begin{aligned} \int_0^m f(x)dx &= \sum_0^{m-1} f(\nu) + \sum_1^n L_\nu \Big[\Delta^{\nu-1} f(x) \Big]_0^m + R \\ R &= m L_{n+1} f^{(n+1)}(\xi). \end{aligned}\right\} \qquad (3)$$

We shall refer to this formula as Laplace's formula, although it was probably (apart from the remainder-term) already known by Newton.

The numbers L_ν which, according to (2), are all rational, are easily calculated either directly or by means of differential coefficients of nothing, as from the expansion

$$t^{(\nu)} = \sum_{r=0}^{\nu} \frac{t^r}{r!} D^r 0^{(\nu)} = \sum_{r=0}^{\nu} (-1)^{r+\nu} \frac{t^r}{r!} D^r 0^{(-\nu)}$$

we obtain

$$L_\nu = \frac{(-1)^{\nu+1}}{\nu!} \left[\frac{D0^{(-\nu)}}{2!} - \frac{D^2 0^{(-\nu)}}{3!} + \ldots + (-1)^{\nu+1} \frac{D^\nu 0^{(-\nu)}}{(\nu+1)!} \right]. \qquad (4)$$

These numbers may also be calculated by a recurrence formula, for, from the expansion

$$(1+t)^x = 1 + \frac{x}{1} t + \frac{x^{(2)}}{2!} t^2 + \ldots$$

valid for $|t| < 1$, we get

$$\int_0^1 (1+t)^x dx = \frac{t}{\operatorname{Log}(1+t)} = L_0 + tL_1 + t^2 L_2 + \ldots, \qquad (5)$$

whence

$$t = (L_0 + tL_1 + t^2L_2 + \dots)\left(t - \frac{t^2}{2} + \frac{t^3}{3} - \dots\right),$$

and consequently, on comparing corresponding powers of t on both sides, $L_0 = 1$ and

$$L_\nu = \frac{1}{2}L_{\nu-1} - \frac{1}{3}L_{\nu-2} + \frac{1}{4}L_{\nu-3} - \dots + \frac{(-1)^{\nu+1}}{\nu+1}L_0. \qquad (6)$$

We find for the first few of these coefficients

$$L_0 = 1,\ L_1 = \frac{1}{2},\ L_2 = -\frac{1}{12},\ L_3 = \frac{1}{24},\ L_4 = -\frac{19}{720},$$

$$L_5 = \frac{3}{160},\ L_6 = -\frac{863}{60480},\ L_7 = \frac{275}{24192}.$$

121. If, for $\nu > 2$, we write

$$t^{(\nu)} = -(t-2)(t-3)\dots(t-\nu+1)\cdot t(1-t),$$

we find from (2), by the Theorem of Mean Value,

$$L_\nu = -\frac{(\theta-2)(\theta-3)\dots(\theta-\nu+1)}{\nu!}\int_0^1 t(1-t)dt \qquad (0<\theta<1)$$

or

$$L_\nu = -\frac{1}{6}\cdot\frac{(\theta-2)(\theta-3)\dots(\theta-\nu+1)}{\nu!} \qquad (0<\theta<1). \qquad (7)$$

L_ν has therefore (with the exception of L_0) the sign $(-1)^{\nu-1}$, so that the signs are alternating.

It follows from (7), that

$$|L_\nu| = \frac{1}{6}\cdot\frac{(2-\theta)(3-\theta)\dots(\nu-1-\theta)}{\nu!} \qquad (0<\theta<1), \qquad (8)$$

from which we may conclude that $|L_\nu|$ is comprised between the limits

$$\frac{1}{6\nu(\nu-1)} < |L_\nu| < \frac{1}{6\nu} \qquad (\nu > 2). \qquad (9)$$

L_ν therefore tends to zero, although not very rapidly, when ν tends to infinity. It can be shown that the exact order of magnitude of L_ν for increasing ν is $\frac{1}{\nu \operatorname{Log}^2 \nu}$ and, therefore, essentially closer to $\frac{1}{\nu}$ than to $\frac{1}{\nu^2}$. More precisely speaking, we have[1]

$$L_\nu \backsim \frac{(-1)^{\nu+1}}{\nu \operatorname{Log}^2 \nu}, \tag{10}$$

the sign $\backsim$ denoting that the quotient of the right- and left-hand sides tends to unity for ν tending to infinity.

122. Other summation-formulas of a similar nature may, in principle, be obtained by inserting, in (1), different expansions of $f(t + r)$ in factorials. We confine ourselves to deriving two of these formulas which present advantages in comparison with (3).

We obtain from §4 (20)

$$\tfrac{1}{2}[f(n+x) + f(n-x)] = \sum_{\nu=0}^{r-1} \frac{x^{2\nu}}{(2\nu)!} \delta^{2\nu} f(n) + \frac{x^{[2r]}}{(2r)!} f^{(2r)}(\xi). \tag{11}$$

If we put

$$K_n = \frac{1}{n!} \int_{-\frac{1}{2}}^{\frac{1}{2}} t^{[n]} dt, \tag{12}$$

whence

$$K_{2\nu+1} = 0, K_{2\nu} = \frac{1}{(2\nu)!} \sum_{s=0}^{\nu} \frac{D^{2s} 0^{[2\nu]}}{2^{2s}(2s+1)!}, \tag{13}$$

we obtain from (11), as the Theorem of Mean Value is applicable to the remainder-term,

$$\int_{-\frac{1}{2}}^{\frac{1}{2}} \tfrac{1}{2} [f(n+x) + f(n-x)]\, dx = \int_{n-\frac{1}{2}}^{n+\frac{1}{2}} f(x)\, dx$$

$$= \sum_{\nu=0}^{r-1} K_{2\nu} \delta^{2\nu} f(n) + K_{2r} f^{(2r)}(\xi)$$

[1] J. F. Steffensen: On Laplace's and Gauss' Summation-Formulas. Skandinavisk Aktuarietidskrift, 1924, p. 2–4.

and hence by summation from $n = 0$ to $n = m - 1$, as $K_0 = 1$,

$$\left.\begin{aligned} \int_{-\frac{1}{2}}^{m-\frac{1}{2}} f(x)dx &= \sum_{0}^{m-1} f(\nu) + \sum_{\nu=1}^{r-1} K_{2\nu}\Big[\delta^{2\nu-1}f(x)\Big]_{-\frac{1}{2}}^{m-\frac{1}{2}} + R \\ R &= m\,K_{2r}\,f^{(2r)}(\xi). \end{aligned}\right\} \quad (14)$$

This formula is called the *first Gaussian summation-formula.* In comparison with (3) it offers the advantage that the terms of even order are missing.

The differences employed in the formula are found in the difference-table on two horizontal lines, starting at $x = -\frac{1}{2}$ and $x = m - \frac{1}{2}$ respectively.

The first few of the constants $K_{2\nu}$ are easily calculated, e.g. by (13), and are

$$K_2 = \frac{1}{24},\ K_4 = -\frac{17}{5760},\ K_6 = \frac{367}{967680},\ K_8 = -\frac{27859}{464486400}.$$

123. From (12) we obtain by the Theorem of Mean Value, in a similar way as in the case of (7),

$$K_{2\nu} = \frac{(-1)^{\nu-1}}{12(2\nu)!}(1-\theta^2)(4-\theta^2)\ldots[(\nu-1)^2-\theta^2] \quad (0 < \theta < \tfrac{1}{2}), \quad (15)$$

showing that these constants have alternating signs and (by putting $\theta = 0$) that

$$|K_{2\nu}| \leq \frac{1}{12\nu^2\binom{2\nu}{\nu}}. \quad (16)$$

The asymptotic expression for $K_{2\nu}$ is obtained more easily than the corresponding one for L_ν. For if we express the factorial in (12) by Gamma-Functions and employ the relation $\Gamma(x)\Gamma(1-x) = \dfrac{\pi}{\sin \pi x}$, we find first

$$K_{2\nu} = (-1)^{\nu-1}\frac{2}{(2\nu)!\,\pi}\int_0^{\frac{1}{2}} \Gamma(\nu+t)\Gamma(\nu-t)t \sin \pi t dt. \quad (17)$$

As $t \sin \pi t$ is positive in the interval, we may apply the Theorem of Mean Value, and get

$$K_{2\nu} = (-1)^{\nu-1} \frac{2\Gamma(\nu+\theta)\Gamma(\nu-\theta)}{(2\nu)!\,\pi^3} \quad (0 < \theta < \tfrac{1}{2}), \tag{18}$$

whence by Stirling's formula

$$K_{2\nu} \backsim \frac{(-1)^{\nu-1}}{2^{2\nu-1}\,\nu^{\frac{3}{2}}\,\pi^{\frac{5}{2}}}. \tag{19}$$

Comparison with the preceding results shows that the $K_{2\nu}$ decrease much more rapidly than the L_ν, so that (14) is also in this respect preferable to Laplace's formula.

124. A formula, analogous to (14), is derived in the following way. We obtain from Bessel's formula §4 (26)

$$\tfrac{1}{2}[f(n+x+\tfrac{1}{2}) + f(n-x+\tfrac{1}{2})] =$$

$$\sum_{\nu=0}^{r-1} \frac{x^{[2\nu+1]-1}}{(2\nu)!} \Box\,\delta^{2\nu} f(n+\tfrac{1}{2}) + \frac{x^{[2r+1]-1}}{(2r)!} f^{(2r)}(\xi). \tag{20}$$

If we put

$$M_n = \frac{1}{n!}\int_{-\frac{1}{2}}^{\frac{1}{2}} t^{[n+1]-1}\,dt, \tag{21}$$

consequently

$$M_{2\nu+1} = 0, M_{2\nu} = \frac{2}{(2\nu)!}\sum_{s=0}^{\nu} \frac{D^{2s+1}0^{[2\nu+1]}}{2^{2s+1}(2s+1)\,(2s+1)!}, \tag{22}$$

we obtain from (20)

$$\int_{-\frac{1}{2}}^{\frac{1}{2}} \tfrac{1}{2}[f(n+x+\tfrac{1}{2}) + f(n-x+\tfrac{1}{2})]\,dx = \int_n^{n+1} f(x)dx$$

$$= \sum_{\nu=0}^{r-1} M_{2\nu}\Box\delta^{2\nu} f(n+\tfrac{1}{2}) + M_{2r} f^{(2r)}(\xi),$$

and from this, by summation from $n = 0$ to $n = m - 1$, and applying the Theorem of Mean Value to the remainder-term,

$$\left.\begin{aligned} \int_0^m f(x)dx &= \sum_0^{m-1} f(\nu) + \tfrac{1}{2}\Big[f(x)\Big]_0^m + \sum_{\nu=1}^{r-1} M_{2\nu}\Big[\Box\delta^{2\nu-1} f(x)\Big]_0^m + R \\ R &= mM_{2r} f^{(2r)}(\xi). \end{aligned}\right\} \tag{23}$$

This formula is called the *second Gaussian summation-formula.* The differences employed are arithmetical means of differences, placed immediately above and below the horizontal lines starting at $x = m$ and $x = 0$.

The first few of the coefficients $M_{2\nu}$ may be calculated, for instance by (22), and are

$$M_2 = -\frac{1}{12}, M_4 = \frac{11}{720}, M_6 = -\frac{191}{60480}, M_8 = \frac{2497}{3628800}.$$

125. We find, from (21), by the Theorem of Mean Value,

$$M_{2\nu} = \frac{(-1)^\nu}{(2\nu)!}(\tfrac{1}{4} - \theta^2)(\tfrac{9}{4} - \theta^2)\ldots\left[\frac{(2\nu-1)^2}{4} - \theta^2\right] \quad (0 < \theta < \tfrac{1}{2}), \quad (24)$$

showing that these constants have alternating signs and (for $\theta = 0$) that

$$|M_{2\nu}| \leq \binom{2\nu}{\nu} 2^{-4\nu} \quad (25)$$

The asymptotic expression for $M_{2\nu}$ is obtained in the same manner as the corresponding one for $K_{2\nu}$. For

$$M_{2\nu} = \frac{2}{(2\nu)!}\int_0^{\frac{1}{2}} \frac{\Gamma(x + \nu + \frac{1}{2})}{\Gamma(x - \nu + \frac{1}{2})}\,dx$$

$$= (-1)^\nu \frac{2}{(2\nu)!\,\pi}\int_0^{\frac{1}{2}} \Gamma(\nu + x + \tfrac{1}{2})\Gamma(\nu - x + \tfrac{1}{2}) \cos \pi x dx$$

$$= (-1)^\nu \frac{2}{(2\nu)!\,\pi^2}\Gamma(\nu + \theta + \tfrac{1}{2})\Gamma(\nu - \theta + \tfrac{1}{2}) \quad (0 < \theta < \tfrac{1}{2}),$$

whence, by Stirling's formula,

$$M_{2\nu} \backsim \frac{(-1)^\nu}{2^{2\nu-1}\nu^{\frac{1}{2}}\pi^{\frac{3}{2}}}. \quad (26)$$

It is seen that the second Gaussian formula also is considerably better than Laplace's formula, although hardly so accurate as the first Gaussian formula.

126. Owing to the form in which we have obtained the remainder-term, we cannot as a rule (an important exception follows in No. 128) see what happens if in (3), (14) and (23) we let m tend to infinity. Cases where the limits of summation and integration are infinite may, however, with approximation, be reduced to cases with finite limits, by stopping at a suitable place, after having ascertained that the neglected parts of sum and integral are comprised within known and sufficiently narrow limits.

In doing so, the error may often be estimated by means of the inequality

$$\sum_{\alpha+1}^{\beta} f(\nu) < \int_{\alpha}^{\beta} f(x)dx < \sum_{\alpha}^{\beta-1} f(\nu), \tag{27}$$

valid for constantly *decreasing* $f(x)$, as appears from a simple geometrical consideration. If $f(x)$ is an increasing function, the theorem may be applied to $-f(x)$. The inequality (27) evidently remains valid if α or β, or both, tend to infinity.

We may also write the inequality in the form of an equation. For, if $A < x < B$, we have

$$x = \tfrac{1}{2}(A+B) + R, \qquad |R| < \tfrac{1}{2}(B-A).$$

If this is applied to (27), we get

$$\left.\begin{aligned} \int_{\alpha}^{\beta} f(x)dx &= \sum_{\alpha}^{\beta-1} f(\nu) - \frac{f(\alpha)-f(\beta)}{2} + R \\ |R| &< \frac{f(\alpha)-f(\beta)}{2}, \end{aligned}\right\} \tag{28}$$

valid for every decreasing and integrable function. If we let $\beta \to \infty$, we obtain in the case of convergence

$$\left.\begin{aligned} \int_{\alpha}^{\infty} f(x)dx &= \sum_{\alpha}^{\infty} f(\nu) - \frac{f(\alpha)}{2} + R \\ |R| &< \frac{f(\alpha)}{2}. \end{aligned}\right\} \tag{29}$$

The formulas (28) and (29) may be considered as primitive summation- or integration-formulas which, however, are often of great use in estimating the values of sums or integrals.

127. As an example, we will calculate $\sum_{1}^{\infty}\frac{1}{\nu^2}$ to 5 decimals, by means of (14). We have, by (29),

$$\sum_{\alpha}^{\infty}\frac{1}{\nu^2}=\frac{1}{\alpha}+\frac{1}{2\alpha^2}+R, \quad |R|<\frac{1}{2\alpha^2}. \tag{30}$$

Putting $\alpha = 500$, we have, therefore,

$$\sum_{1}^{\infty}\frac{1}{\nu^2}=\sum_{1}^{499}\frac{1}{\nu^2}+0.0020020$$

with an error that does not exceed 0.0000020 in absolute value.

As a rule it is advantageous to begin by calculating the first few terms of the sum *directly*, as in that case we can do with fewer terms in (14). We find by a table of reciprocals to seven places of decimals

$$\sum_{1}^{15}\frac{1}{\nu^2}=1.5804403$$

with an error that does not exceed 0.0000002; consequently

$$\sum_{1}^{\infty}\frac{1}{\nu^2}=\sum_{16}^{499}\frac{1}{\nu^2}+1.5824423$$

with an error that does not exceed 0.0000022.

We may now calculate $\sum_{16}^{499}\frac{1}{\nu^2}$ by (14). Putting, in this formula, $r = 3$, we find for the remainder-term

$$R=484\,K_6\cdot\frac{7!}{\xi^8}.$$

The difference of highest order retained being δ^3, we obtain an upper limit to R by taking $\xi = 14$; consequently

$$|R|<484\cdot\frac{367}{967680}\cdot\frac{5040}{14^8}=0.00000063.$$

We may therefore, with an error that does not exceed one unit in the sixth place, put

$$\sum_{16}^{499} \frac{1}{\nu^2} = \int_{15.5}^{499.5} \frac{dx}{x^2} - \frac{1}{24}\left[\delta \frac{1}{x^2}\right]_{15.5}^{499.5} + \frac{17}{5760}\left[\delta^3 \frac{1}{x^2}\right]_{15.5}^{499.5}.$$

Now it is easy to convince oneself that the values of δ and δ^3 at the upper limit are negligible in consideration of the degree of accuracy required in the result. We may therefore put

$$\sum_{16}^{499} \frac{1}{\nu^2} = \frac{1}{15.5} - \frac{1}{499.5} + \frac{1}{24}\,\delta\,\frac{1}{15.5^2} - \frac{17}{5760}\,\delta^3\,\frac{1}{15.5^2}.$$

If we form the table

x	$\frac{1}{x^2}$	δ	δ^2	δ^3
14	0.0051020			
		−6576		
15	44444		1194	
		−5382		−272
16	39062		922	
		−4460		
17	34602			

we find

$$\sum_{16}^{499} \frac{1}{\nu^2} = \frac{1}{15.5} - \frac{1}{499.5} - \frac{0.0005382}{24} + \frac{17}{5760} \cdot 0.0000272$$

$$= 0.0624918$$

and finally with an error which is certainly smaller than one half unit of the fifth decimal

$$\sum_{1}^{\infty} \frac{1}{\nu^2} = 1.5804403 + 0.0624918 + 0.0020020$$

$$= 1.644934.$$

We shall see later on (§13(38)), that

$$\sum_{1}^{\infty} \frac{1}{\nu^2} = \frac{\pi^2}{6} = 1.644934,$$

so that we have, as a matter of fact, obtained six correct decimals, although the method, on account of the inaccuracy of (30), only guarantees five.

Formula (30) may be considered as an expression for the remainder-term in the convergent series $\sum_{1}^{\infty} \frac{1}{\nu^2}$, if we stop at $\alpha - 1$ terms. The question may be asked, how many terms of this series are required for the *direct* calculation of the sum of the series to five places of decimals. We must then determine α in such a way that

$$\frac{1}{\alpha} + \frac{1}{2\,\alpha^2} + R < \frac{1}{200000},$$

where $|R| < \frac{1}{2\alpha^2}$; but it follows that $\frac{1}{\alpha} < \frac{1}{200000}$, so that $\alpha > 200000$. It would, therefore, be necessary to calculate and, thereafter, add up, more than 200000 terms in order to obtain five correct decimals. As each of these terms may contain an error up to one half unit of the last decimal, it would, moreover, be necessary to calculate each term to 11 decimals, in order to secure 5 decimals in the sum. This is a striking example of the uselessness o convergence, in the mathematical sense of the word, for numerica calculations.

128. The Error-Test is immediately applicable to (3), (14) and (23). Assuming that the first neglected term does not vanish, the conditions are: in the case of Laplace's formula, that $f^{(n+1)}$ and $f^{(n+2)}$ keep their signs and have the same sign; in the case of the two Gaussian formulas, that $f^{(2r)}$ and $f^{(2r+2)}$ keep their signs and have the same sign. In these cases the error is, therefore, less than the first neglected term and has the same sign.

If the Error-Test is applicable, it remains applicable for m tending to infinity. If the integrals and sums in (3), (14) and (23) exist for $m \to \infty$, the differences employed in these formulas must vanish for $m \to \infty$ and R exist for $m \to \infty$, and even if we do not know the form of the remainder-term, we may apply the Error-Test. Under these circumstances the three formulas may be written

$$\int_0^\infty f(x)dx \doteq \sum_0^\infty f(\nu) - \sum_1^n L_\nu \triangle^{\nu-1} f(0) + R,$$

$$\int_{-\frac{1}{2}}^\infty f(x)dx = \sum_0^\infty f(\nu) - \sum_{\nu=1}^{r-1} K_{2\nu}\delta^{2\nu-1} f(-\tfrac{1}{2}) + R,$$

$$\int_0^\infty f(x)dx = \sum_0^\infty f(\nu) - \tfrac{1}{2}f(0) - \sum_{\nu=1}^{r-1} M_{2\nu}\square\,\delta^{2\nu-1} f(0) + R.$$

If the Error-Test is applicable, these formulas are preferable to the method employed in the numerical example. It is easily seen that the conditions are satisfied in this very case, so that we may put

$$\sum_{16}^\infty \frac{1}{\nu^2} = \frac{1}{15.5} + K_2\delta\frac{1}{15.5^2} + K_4\delta^3\frac{1}{15.5^2}$$

$$= \frac{1}{15.5} - \frac{0.0005382}{24} + \frac{17}{5760}0.0000272 = 0.0644938,$$

the error not exceeding $K_6\delta^5\dfrac{1}{15.5^2}$. As, in this case, it is easier to calculate $f^{(5)}$ than δ^5, we prefer, however, as upper limit to the error, to take

$$|K_6 f^{(5)}(\xi)| \leq \frac{367}{967680}\cdot\frac{720}{13^7} < 0.0000000044,$$

so that the figures retained in $\sum_{16}^\infty \dfrac{1}{\nu^2}$ are reliable. If, to this, we add the directly calculated value $\sum_1^{15}\dfrac{1}{\nu^2} = 1.5804403$, we get $\sum_1^\infty \dfrac{1}{\nu^2} =$ 1.6449341, where the correctness of the first 6 decimals is secured (the last decimal happens also to be correct).

129. *Repeated integrals* may, as shown in No. 117, be reduced to single integrals. Thus, putting, in §11(29), $\alpha = -\frac{1}{2}$, $\beta = m - \frac{1}{2}$, $k = 1$, we get

$$\int_{-\frac{1}{2}}^{m-\frac{1}{2}}\int_{-\frac{1}{2}}^{x} F(x)dx^2 = \int_{-\frac{1}{2}}^{m-\frac{1}{2}} (m - \tfrac{1}{2} - x)F(x)dx. \qquad (31)$$

We need therefore only, in the first Gaussian formula, put

$$f(x) = (m - \tfrac{1}{2} - x)\, F(x). \tag{32}$$

The matter might be left at that; but we may also transform the formula in such a way that the terms on the right become sums and differences of $F(x)$ instead of $f(x)$. In this way we avoid forming a table of $f(x)$ and can do with the given table of $F(x)$. The process is as follows.

We apply §2 (18) to (32) and find

$$\delta^{2\nu-1} f(x) = (m + \nu - 1 - x)\, \delta^{2\nu-1} F(x) - (2\nu - 1)\, \delta^{2\nu-2} F(x + \tfrac{1}{2});$$

but

$$F(x + \tfrac{1}{2}) = E^{\frac{1}{2}} F(x) = \left(\Box + \frac{\delta}{2}\right) F(x),$$

so that

$$\delta^{2\nu-1} f(x) = (m - x - \tfrac{1}{2})\, \delta^{2\nu-1} F(x) - (2\nu - 1)\, \Box\, \delta^{2\nu-2} F(x),$$

and hence

$$\Big[\delta^{2\nu-1} f(x)\Big]_{-\frac{1}{2}}^{m-\frac{1}{2}} = -m\delta^{2\nu-1} F(-\tfrac{1}{2}) - (2\nu-1) \Big[\Box\, \delta^{2\nu-2} F(x)\Big]_{-\frac{1}{2}}^{m-\frac{1}{2}}. \tag{33}$$

Further, we have

$$\sum_{\nu=0}^{m-1} f(\nu) = \sum_{\nu=0}^{m-1} (m - 1 - \nu)\, F(\nu) + \tfrac{1}{2} \sum_{\nu=0}^{m-1} F(\nu)$$

or

$$\sum_{\nu=0}^{m-1} f(\nu) = \Sigma^2 F(m) + \tfrac{1}{2}\, \Sigma F(m); \tag{34}$$

the sums $\Sigma\, F(m)$ and $\Sigma^2 F(m)$ are, in the sum-table for summation from the top, found on the upward sloping line, commencing at $F(m)$.

Finally, by Leibnitz' theorem

$$f^{(2r)}(\xi) = (m - \tfrac{1}{2} - \xi)\, F^{(2r)}(\xi) - 2r\, F^{(2r-1)}(\xi). \tag{35}$$

If, now, we insert (32)–(35) in (14), we find

$$\left.\begin{aligned}\int_{-\frac{1}{2}}^{m-\frac{1}{2}}\int_{-\frac{1}{2}}^{x} F(x)dx^2 = \Sigma^2 F(m) + \tfrac{1}{2}\Sigma F(m) - m\sum_{\nu=1}^{r-1} K_{2\nu}\,\delta^{2\nu-1}F(-\tfrac{1}{2})\\ - \sum_{\nu=1}^{r-1} J_{2\nu}\Big[\Box\,\delta^{2\nu-2}F(x)\Big]_{-\frac{1}{2}}^{m-\frac{1}{2}} + R\end{aligned}\right\}\tag{36}$$

where

$$J_{2\nu} = (2\nu - 1)\,K_{2\nu}$$

and

$$R = mK_{2r}\,[(m - \tfrac{1}{2} - \xi)\,F^{(2r)}(\xi) - 2r\,F^{(2r-1)}(\xi)]. \tag{37}$$

This is the required formula. The first few of the coefficients $J_{2\nu}$ are

$$J_2 = \frac{1}{24},\ J_4 = -\frac{17}{1920},\ J_6 = \frac{367}{193536},\ J_8 = -\frac{27859}{66355200}.$$

130. An analogous formula for repeated integration is, in a similar way, derived from the second Gaussian formula. We find, putting in §11 (29) $\alpha = 0, \beta = m, k = 1$,

$$\int_0^m\int_0^x F(x)\,dx^2 = \int_0^m (m - x)\,F(x)\,dx \tag{38}$$

and therefore put, in (23),

$$f(x) = (m - x)\,F(x). \tag{39}$$

Now we have, by §2 (18),

$$\delta^{2\nu-1}f(x) = (m + \nu - \tfrac{1}{2} - x)\,\delta^{2\nu-1}F(x) - (2\nu - 1)\,\delta^{2\nu-2}F(x + \tfrac{1}{2}),$$

and hence

$$\begin{aligned}\Box\,\delta^{2\nu-1}f(x) &= (m + \nu - \tfrac{1}{2})\,\Box\,\delta^{2\nu-1}F(x) - \tfrac{1}{2}\,(x + \tfrac{1}{2})\,\delta^{2\nu-1}F(x + \tfrac{1}{2})\\ &\quad - \tfrac{1}{2}\,(x - \tfrac{1}{2})\,\delta^{2\nu-1}F(x - \tfrac{1}{2}) - (2\nu - 1)\,\Box\,\delta^{2\nu-2}F(x + \tfrac{1}{2})\\ &= (m+\nu-\tfrac{1}{2})\,\Box\,\delta^{2\nu-1}F(x) - x\,\Box\,\delta^{2\nu-1}F(x) - \tfrac{1}{4}\delta^{2\nu}F(x) - (\nu-\tfrac{1}{2})\,(E+1)\,\delta^{2\nu-2}F(x)\end{aligned}$$

or, as

$$E = 1 + \Box\,\delta + \tfrac{1}{2}\delta^2,$$

and leaving out $F(x)$,

$$= (m-x)\,\Box\,\delta^{2\nu-1} + (\nu-\tfrac{1}{2})\,\Box\,\delta^{2\nu-1} - \tfrac{1}{4}\delta^{2\nu} - (\nu-\tfrac{1}{2})\,(2+\Box\delta+\tfrac{1}{2}\delta^2)\,\delta^{2\nu-2}$$

$$= (m-x)\,\Box\,\delta^{2\nu-1} - (2\nu-1)\,\delta^{2\nu-2} - \tfrac{\nu}{2}\,\delta^{2\nu},$$

so that

$$\Big[\Box\,\delta^{2\nu-1}f(x)\Big]_0^m = -m\Box\,\delta^{2\nu-1}F(0) - (2\nu-1)\Big[\delta^{2\nu-2}F(x)\Big]_0^m - \frac{\nu}{2}\Big[\delta^{2\nu}F(x)\Big]_0^m. \tag{40}$$

Further, we have

$$\sum_{0}^{m-1} f(\nu) = \sum_{\nu=0}^{m-1} (m-\nu)\,F(\nu) = \Sigma^2 F(m+1) \tag{41}$$

and finally

$$f^{(2r)}(\xi) = (m-\xi)\,F^{(2r)}(\xi) - 2r\,F^{(2r-1)}(\xi). \tag{42}$$

We now insert (39)–(42) in (23), putting

$$N_{2\nu} = (2\nu-1)\,M_{2\nu} + \frac{\nu-1}{2}M_{2\nu-2}; \tag{43}$$

and a simple rearrangement of the terms leads to the formula

$$\left.\begin{aligned} \int_0^m\!\!\int_0^x F(x)dx^2 &= \Sigma^2F(m+1) - m\Big[\tfrac{1}{2}F(0) + \sum_{\nu=1}^{r-1} M_{2\nu}\Box\,\delta^{2\nu-1}F(0)\Big] \\ &- \sum_{\nu=1}^{r-1} N_{2\nu}\Big[\delta^{2\nu-2}F(x)\Big]_0^m - \frac{r-1}{2}M_{2r-2}\Big[\delta^{2r-2}F(x)\Big]_0^m + R \end{aligned}\right\} \tag{44}$$

where

$$R = mM_{2r}[(m-\xi)F^{(2r)}(\xi) - 2r\,F^{(2r-1)}(\xi)]. \tag{45}$$

The first few of the coefficients $N_{2\nu}$ are

$$N_2 = -\frac{1}{12},\ N_4 = \frac{1}{240},\ N_6 = -\frac{31}{60480},\ N_8 = \frac{289}{3628800}.$$

131. The formulas (36) and (44) are closely related to the formulas used (without remainder-term) by astronomers for the numerical integration of differential equations.[1]

[1] Steffensen: l. c., p. 12–15.

§13. Bernoulli's Polynomials

132. We shall frequently have to make use of certain polynomials which are related to the factorials and called Bernoulli's polynomials. These polynomials are of great importance in mathematical analysis, and a considerable amount of literature exists about them. Here, however, we confine ourselves to deriving those of their properties which will be of use to us later on. Bernoulli's polynomials are by various authors defined in slightly different ways; the definition preferred here is equivalent to that of Nörlund.[1]

133. Bernoulli's polynomial of degree m is denoted by $B_m(x)$ and is defined as the polynomial that satisfies at the same time the two relations

$$\Delta B_m(x) = m x^{m-1} \tag{1}$$

and

$$D B_m(x) = m B_{m-1}(x). \tag{2}$$

It is obvious that a polynomial with such simple properties must have important applications. To begin with, we proceed to show that the polynomial is perfectly determined by the two equations (1) and (2).

If the polynomial exists, $B_m(x + h)$ will also be a polynomial of degree m, and this polynomial admits, by Taylor's theorem, the unique expansion

$$B_m(x + h) = \sum_{\nu=0}^{m} \frac{h^\nu}{\nu!} D^\nu B_m(x).$$

But this may, by (2), be written

$$B_m(x + h) = \sum_{\nu=0}^{m} \binom{m}{\nu} h^\nu B_{m-\nu}(x)$$

or, writing $m - \nu$ for ν,

$$B_m(x + h) = \sum_{\nu=0}^{m} \binom{m}{\nu} h^{m-\nu} B_\nu(x). \tag{3}$$

[1] N. E. Nörlund: Differenzenrechnung, p. 17–23; Mémoire sur les polynomes de Bernoulli, Acta Mathematica, vol. 43, p. 121. See also J. F. Steffensen: On a generalization of Nörlund's Polynomials, Bulletin de l'Académie Royale de Danemark, 1926, vii, 5.

If, in this, we put $h = 1$, we find, by (1),

$$\sum_{\nu=0}^{m-1} \binom{m}{\nu} B_\nu(x) = mx^{m-1}. \tag{4}$$

By this formula, the polynomials $B_\nu(x)$ are perfectly determined; we find

$$\left.\begin{aligned}
B_0(x) &= 1 \\
B_1(x) &= x - \tfrac{1}{2} \\
B_2(x) &= x^2 - x + \tfrac{1}{6} \\
B_3(x) &= x^3 - \tfrac{3}{2}x^2 + \tfrac{1}{2}x \\
B_4(x) &= x^4 - 2x^3 + x^2 - \tfrac{1}{30} \\
B_5(x) &= x^5 - \tfrac{5}{2}x^4 + \tfrac{5}{3}x^3 - \tfrac{1}{6}x \\
B_6(x) &= x^6 - 3x^5 + \tfrac{5}{2}x^4 - \tfrac{1}{2}x^2 + \tfrac{1}{42} \\
B_7(x) &= x^7 - \tfrac{7}{2}x^6 + \tfrac{7}{2}x^5 - \tfrac{7}{6}x^3 + \tfrac{1}{6}x \\
B_8(x) &= x^8 - 4x^7 + \tfrac{14}{3}x^6 - \tfrac{7}{3}x^4 + \tfrac{2}{3}x^2 - \tfrac{1}{30} \\
B_9(x) &= x^9 - \tfrac{9}{2}x^8 + 6x^7 - \tfrac{21}{5}x^5 + 2x^3 - \tfrac{3}{10}x \\
B_{10}(x) &= x^{10} - 5x^9 + \tfrac{15}{2}x^8 - 7x^6 + 5x^4 - \tfrac{3}{2}x^2 + \tfrac{5}{66}.
\end{aligned}\right\} \tag{5}$$

134. The values of $B_\nu(x)$ for $x = 0$ are called Bernoulli's numbers and are denoted by B_ν; that is, $B_\nu = B_\nu(0)$. The first few of these values may be taken from (5); but they may also be treated independently, as we find, by (4), for $x = 0$

$$B_0 = 1; \quad \sum_{\nu=0}^{m} \binom{m}{\nu} B_\nu = B_m \qquad (m > 1). \tag{6}$$

This relation may be written in a convenient symbolical form as

$$(B + 1)^m - B^m = 0 \qquad (m > 1). \tag{7}$$

The meaning of this relation is that we develop by the binomial theorem, and thereafter replace B^ν by B_ν.

The first few of Bernoulli's numbers are (leaving out zero values)

$$B_0 = 1,\ B_1 = -\frac{1}{2},\ B_2 = \frac{1}{6},\ B_4 = -\frac{1}{30},\ B_6 = \frac{1}{42},$$

$$B_8 = -\frac{1}{30},\ B_{10} = \frac{5}{66},\ B_{12} = -\frac{691}{2730},\ B_{14} = \frac{7}{6},\ B_{16} = -\frac{3617}{510}.$$

Bernoulli's numbers being known, the values of $B_\nu(1)$ are found from (1), putting $x = 0$; we get

$$B_1(1) = B_1 + 1; \qquad B_\nu(1) = B_\nu \qquad (\nu \neq 1). \tag{8}$$

Bernoulli's polynomials may be expressed explicitly by Bernoulli's numbers. If, in (3), we put $x = 0$ and afterwards replace h by x, we find

$$B_m(x) = \sum_{\nu=0}^{m} \binom{m}{\nu} B_\nu x^{m-\nu} \tag{9}$$

which is the required expression. This may also be written in the symbolical form

$$B_m(x) = (x + B)^m. \tag{10}$$

135. Bernoulli's polynomials and numbers satisfy a great many relations which are most easily obtained in symbolical form. We find first from (7), multiplying by c_m and summing from $m = 2$ to $m = n$

$$\sum_{2}^{n} c_m (B + 1)^m - \sum_{2}^{n} c_m B^m = 0$$

and hence, $P(x)$ denoting the arbitrary polynomial $P(x) = \sum_{0}^{n} c_m x^m$,

$$P(B + 1) - P(B) = c_1,$$

or

$$P(B + 1) - P(B) = P'(0). \tag{11}$$

If, in this relation, instead of $P(t)$ we write $P(t + x)$, we obtain the more general symbolical relation

$$P(x + B + 1) - P(x + B) = P'(x). \tag{12}$$

The difference-equation

$$\Delta f(x) = P'(x) \tag{13}$$

therefore has the solution

$$\left.\begin{aligned} f(x) &= P(x+B) \\ &= P(x) + \frac{B_1}{1!}P'(x) + \frac{B_2}{2!}P''(x) + \dots, \end{aligned}\right\} \quad (14)$$

the number of terms being finite.

If, in particular, we put $P(x) = (x+h)^m$, we find

$$B_m(x+h) = (x+B+h)^m \quad (15)$$

or the symbolical form of (3), as is seen by expanding the right-hand side of (15) in powers of h and employing (10).

If, in (14), we replace x by $x+h$, we have the symbolical relation

$$f(x+h) = P(x+h+B) = P(x+B(h)),$$

the meaning of which is that after having expanded in powers of $B(h)$, we must replace $B^\nu(h)$ by $B_\nu(h)$, that is

$$\left.\begin{aligned} f(x+h) &= P(x+B(h)) \\ &= P(x) + \frac{B_1(h)}{1!}P'(x) + \frac{B_2(h)}{2!}P''(x) + \dots \end{aligned}\right\} \quad (16)$$

Performing the operation Δ on both sides of this equation, we find, by (13),

$$P'(x+h) = \Delta P(x) + \frac{B_1(h)}{1!}\Delta P'(x) + \frac{B_2(h)}{2!}\Delta P''(x) + \dots, \quad (17)$$

a relation which is called *Euler-Maclaurin's formula for a polynomial.*

The relation (12) may evidently be written in the form

$$P(B(x)+1) - P(B(x)) = P'(x); \quad (18)$$

by suitable choice of the polynomial P we may obtain any number of special relations between Bernoullian polynomials. We shall, however, not go into this, but proceed to derive certain properties which are more important for the applications we are going to make of these polynomials.

136. The difference-equation

$$f(x+1) - f(x) = (\nu + 1)x^\nu$$

is satisfied by $B_{\nu+1}(x)$ and also by $(-1)^{\nu+1}B_{\nu+1}(1-x)$, as is seen, if in the equation

$$B_{\nu+1}(x+1) - B_{\nu+1}(x) = (\nu+1)x^\nu$$

we replace x by $-x$. We have, therefore,

$$B_{\nu+1}(x) - (-1)^{\nu+1}B_{\nu+1}(1-x) = \text{const.},$$

whence, differentiating and making use of (2),

$$B_\nu(1-x) = (-1)^\nu B_\nu(x). \tag{19}$$

It is seen that $B_{2r}(x)$ *is symmetrical about the point* $x = \frac{1}{2}$, and that $B_{2r+1}(\frac{1}{2}) = 0$. Further, we obtain for $x = 0$ the relation $B_\nu(1) = (-1)^\nu B_\nu$ which, jointly with (8), shows that $B_{2r+1} = 0$ for $r > 0$.

137. It can be shown that

$$B_\nu(\tfrac{1}{2}) = (2^{1-\nu} - 1)B_\nu. \tag{20}$$

This relation is obviously true for $\nu = 0$ and $\nu = 1$, so that we may assume $\nu > 1$. We obtain from (4) for $x = \frac{1}{2}$

$$\sum_{\nu=0}^{m-1} \binom{m}{\nu} B_\nu(\tfrac{1}{2}) = 2^{1-m}m.$$

This relation determines $B_\nu(\frac{1}{2})$ completely, and we proceed to show that it is satisfied by (20). Inserting this expression, making use of (6) and (9), and assuming $m > 1$, we find, in fact,

$$\sum_{\nu=0}^{m-1} \binom{m}{\nu}(2^{1-\nu} - 1)B_\nu = \sum_{\nu=0}^{m-1} \binom{m}{\nu} 2^{1-\nu}B_\nu$$

$$= \sum_{\nu=0}^{m} \binom{m}{\nu} 2^{1-\nu}B_\nu - 2^{1-m}B_m$$

$$= 2^{1-m}B_m(2) - 2^{1-m}B_m$$

$$= 2^{1-m}(B_m(1) + m - B_m) = 2^{1-m}m.$$

138. An important formula is obtained if, from (1), we derive

$$(x+s)^m = \frac{1}{m+1}\triangle B_{m+1}(x+s)$$

and sum on both sides from $s = 0$ to $s = n - 1$. We find for $m > 0$

$$\sum_{s=0}^{n-1} (x+s)^m = \frac{1}{m+1}[B_{m+1}(x+n) - B_{m+1}(x)]$$

or

$$\left.\begin{aligned}\sum_{s=0}^{n-1} (x+s)^m &= \frac{1}{m+1}\triangle_n B_{m+1}(x) \\ &= \int_0^n B_m(x+t)dt.\end{aligned}\qquad (m > 0)\right\} \qquad (21)$$

In particular, we have for $x = 0$ and $x = \dfrac{1}{2}$

$$\sum_{s=1}^{n-1} s^m = \frac{1}{m+1}\triangle_n B_{m+1}(0) \qquad (m > 0), \qquad (22)$$

$$\sum_{s=0}^{n-1} (s+\tfrac{1}{2})^m = \frac{1}{m+1}\triangle_n B_{m+1}(\tfrac{1}{2}), \qquad (23)$$

the latter formula being also valid for $m = 0$.

Formula (22) expresses the fundamental property of Bernoulli's polynomials, and is the one that first established their importance in mathematical analysis. It gives a new solution of a summation-problem which we have already, in §10, solved by other means.

139. In order to derive the other properties of Bernoulli's polynomials, we will first express these in a different way which will be of use to us later on.

We proceed to prove that

$$\triangle_x B_{2r}(0) \equiv B_{2r}(x) - B_{2r}$$

can be expanded in powers of

$$z = x - x^2,$$

and that the coefficients in this expansion all have the same sign (Jacobi's theorem).[1]

We put

$$(-1)^r B_{2r}(x) = \sum_{\nu=0}^{r} G_\nu^{(r)} z^{r-\nu} \tag{24}$$

and are first going to prove that the coefficients can be determined.

If we differentiate (24) twice with respect to x, and note that

$$D_x^2 z^k = k^{(2)} z^{k-2} - (2k)^{(2)} z^{k-1},$$

we obtain

$$(-1)^r (2r)^{(2)} B_{2r-2}(x) = \sum_{\nu=0}^{r} G_\nu^{(r)} [(r-\nu)^{(2)} z^{r-\nu-2} - (2r-2\nu)^{(2)} z^{r-\nu-1}]$$

$$= \sum_{\nu=1}^{r-1} [(r-\nu+1)^{(2)} G_{\nu-1}^{(r)} - (2r-2\nu)^{(2)} G_\nu^{(r)}] z^{r-\nu-1} - (2r)^{(2)} G_0^{(r)} z^{r-1}.$$

But if we compare this expansion with

$$(-1)^{r-1} B_{2r-2}(x) = \sum_{\nu=0}^{r-1} G_\nu^{(r-1)} z^{r-\nu-1},$$

resulting from (24), if we write $r - 1$ for r, we find

$$(2r - 2\nu)^{(2)} G_\nu^{(r)} = (2r)^{(2)} G_\nu^{(r-1)} + (r - \nu + 1)^{(2)} G_{\nu-1}^{(r)}. \tag{25}$$

This is a recurrence formula for the calculation of the coefficients $G_\nu^{(r)}$. The necessary initial values are obtained by noting that

$$G_0^{(r)} = 1; \qquad G_{r-1}^{(r)} = 0 \quad (r > 1). \tag{26}$$

The former of these values is found by observing that the coefficient of x^{2r} in $B_{2r}(x)$ is 1; the latter follows from the fact, that for $r > 1$, $B_{2r}(x)$ does not contain a term of the first degree in x, as appears from (9).

If, finally, we note that

$$G_r^{(r)} = (-1)^r B_{2r} \tag{27}$$

[1] Journal für die reine und angewandte Mathematik (Crelle's Journal), vol. 12, p. 268–9.

(as results from (24) for $x = 0$), we may write (24), for $r > 1$,

$$\Delta_x B_{2r}(0) = (-1)^r \sum_{\nu=0}^{r-2} G_\nu^{(r)} (x - x^2)^{r-\nu} \quad (r > 1), \tag{28}$$

the coefficients $G_\nu^{(r)}$ being, according to (25) and (26), all *positive.*

We give below a table of $G_\nu^{(r)}$ which is sufficient for most practical purposes.

Table of $G_\nu^{(r)}$

r	$\nu = 0$	1	2	3	4	5	6	7	8
2	1	0							
3	1	$\frac{1}{2}$	0						
4	1	$\frac{4}{3}$	$\frac{2}{3}$	0					
5	1	$\frac{5}{2}$	3	$\frac{3}{2}$	0				
6	1	4	$\frac{17}{2}$	10	5	0			
7	1	$\frac{35}{6}$	$\frac{287}{15}$	$\frac{118}{3}$	$\frac{691}{15}$	$\frac{691}{30}$	0		
8	1	8	$\frac{112}{3}$	$\frac{352}{3}$	$\frac{718}{3}$	280	140	0	
9	1	$\frac{21}{2}$	66	293	$\frac{4557}{5}$	$\frac{3711}{2}$	$\frac{10851}{5}$	$\frac{10851}{10}$	0

It was calculated by first filling up the places where the values, according to (26), are 1 and 0, and thereafter applying (25). But if Bernoulli's numbers are already considered to be known, the calculation may be simplified by observing that

$$G_{r-2}^{(r)} = (-1)^r r(2r-1) B_{2r-2}; \quad G_{r-3}^{(r)} = 2G_{r-2}^{(r)}. \tag{29}$$

These relations are obtained from (25) for $\nu = r - 1$ and $\nu = r - 2$ respectively.

By means of (29), we may first fill up the diagonal of the table, and after that, the first parallel to the diagonal, sloping down to the right.

It results from the first relation (29), that $(-1)^{r+1} B_{2r}$ is, for $r > 0$, always a *positive* number.

By means of the table of $G_\nu^{(r)}$ we may immediately write down the Bernoullian polynomials of even degree, e.g.

$$\Delta_x B_{10}(0) = -(z^5 + \tfrac{5}{2}z^4 + 3z^3 + \tfrac{3}{2}z^2).$$

It is seen how much shorter this expression is than the corresponding one by (5). In practical calculations with Bernoulli's polynomials, the Jacobian form is, therefore, nearly always preferable.

140. The polynomials of odd degree may be treated in a similar way, most simply by differentiating (28). If we put

$$H_\nu^{(r)} = \frac{r-\nu}{2r} G_\nu^{(r)}, \tag{30}$$

we get

$$B_{2r-1}(x) = (-1)^r (1-2x) \sum_{\nu=0}^{r-2} H_\nu^{(r)} (x-x^2)^{r-\nu-1} \quad (r > 1). \tag{31}$$

The coefficients $H_\nu^{(r)}$ are evidently also positive. By means of the table of $H_\nu^{(r)}$ given below, we may, for instance, immediately write down

$$B_9(x) = -(1-2x)(\tfrac{1}{2}z^4 + z^3 + \tfrac{9}{10}z^2 + \tfrac{3}{10}z).$$

Table of $H_\nu^{(r)}$

r	$\nu = 0$	1	2	3	4	5	6	7	8
2	$\frac{1}{2}$	0							
3	$\frac{1}{2}$	$\frac{1}{6}$	0						
4	$\frac{1}{2}$	$\frac{1}{2}$	$\frac{1}{6}$	0					
5	$\frac{1}{2}$	1	$\frac{9}{10}$	$\frac{3}{10}$	0				
6	$\frac{1}{2}$	$\frac{5}{3}$	$\frac{17}{6}$	$\frac{5}{2}$	$\frac{5}{6}$	0			
7	$\frac{1}{2}$	$\frac{5}{2}$	$\frac{41}{6}$	$\frac{236}{21}$	$\frac{691}{70}$	$\frac{691}{210}$	0		
8	$\frac{1}{2}$	$\frac{7}{2}$	14	$\frac{110}{3}$	$\frac{359}{6}$	$\frac{105}{2}$	$\frac{35}{2}$	0	
9	$\frac{1}{2}$	$\frac{14}{3}$	$\frac{77}{3}$	$\frac{293}{3}$	$\frac{1519}{6}$	$\frac{1237}{3}$	$\frac{3617}{10}$	$\frac{3617}{30}$	0

141. As $G_\nu^{(r)}$ and $H_\nu^{(r)}$ are positive, we may, from (28) and (31) draw various conclusions concerning the polynomials $\Delta_x B_{2r}(0)$ and $B_{2r-1}(x)$. For instance, (28) shows that $\Delta_x B_{2r}(0)$ does not change its sign in the interval (0, 1), and that this sign is $(-1)^r$. This also holds for $r = 1$. It is also seen that this polynomial contains, for $r > 1$, the factor $x^2(x-1)^2$ and has, therefore, double roots at $x = 0$ and $x = 1$. In the interval (0, 1), $\Delta_x B_{2r}(0)$ has a numerically largest value which is attained for $x = \frac{1}{2}$; this also holds for $r = 1$, and the polynomial is symmetrical about this point.

By (31) it is seen that $B_{2r-1}(x)$ contains, for $r > 1$, the factor $x(x-1)(x-\frac{1}{2})$, and that 0, $\frac{1}{2}$ and 1 are single roots.

142. Bernoulli's polynomials may be defined in another way which is useful on many occasions. It may be proved that, for $|\tau| < 2\pi$,

$$\frac{\tau e^{\tau x}}{e^\tau - 1} = \sum_0^\infty \frac{\tau^\nu}{\nu!} B_\nu(x). \tag{32}$$

For, multiplying both sides by $e^\tau - 1$ and expanding, we get

$$\tau + \frac{\tau^2}{1!}x + \frac{\tau^3}{2!}x^2 + \ldots =$$

$$\left(\tau + \frac{\tau^2}{2!} + \frac{\tau^3}{3!} + \ldots\right)\left(B_0(x) + \frac{\tau}{1!}B_1(x) + \frac{\tau^2}{2!}B_2(x) + \ldots\right).$$

If, now, we compare the coefficient of τ^m on both sides, we arrive at the recurrence formula (4) which, as we have seen, suffices for determining these polynomials.

If, in particular, we put, in (32), $x = 0$, we find

$$\frac{\tau}{e^\tau - 1} = \sum_0^\infty \frac{\tau^\nu}{\nu!} B_\nu. \tag{33}$$

For $x = \frac{1}{2}$, we get

$$\frac{\tau}{e^{\frac{\tau}{2}} - e^{-\frac{\tau}{2}}} = \sum_0^\infty \frac{\tau^\nu}{\nu!} B_\nu(\tfrac{1}{2}). \tag{34}$$

143. We will, finally, derive an expression for $B_{2\nu}$, showing how these numbers behave for large values of ν. We start from the well-known expansion.[1]

$$\frac{x}{e^x - 1} = 1 - \frac{x}{2} + 2 \sum_{\nu=1}^{\infty} \frac{x^2}{x^2 + 4\pi^2\nu^2}. \tag{35}$$

If we put

$$s_n = \sum_{\nu=1}^{\infty} \frac{1}{\nu^n} \tag{36}$$

and note that

$$\frac{x^2}{x^2 + 4\pi^2\nu^2} = \sum_{n=1}^{\infty} (-1)^{n+1} \left(\frac{x}{2\pi\nu}\right)^{2n},$$

we obtain from (35), changing the order of summation,

$$\frac{x}{e^x - 1} = 1 - \frac{x}{2} + \sum_{n=1}^{\infty} (-1)^{n+1} \frac{2s_{2n}}{(2\pi)^{2n}} x^{2n}. \tag{37}$$

By comparison with (33), we therefore learn that

$$B_{2\nu} = (-1)^{\nu+1} \frac{2(2\nu)!}{(2\pi)^{2\nu}} s_{2\nu} \quad (\nu > 0). \tag{38}$$

As $s_{2\nu} \to 1$ for $\nu \to \infty$, it is seen that Bernoulli's numbers which, to begin with, have small values, tend to infinity with ν. The exact order of magnitude can, if desired, be found by Stirling's theorem.

§14. Euler's Summation-Formula

144. The relations between sums and integrals derived in §12 were of such a form that the corrective terms were expressed as differences of the function under consideration. Now we shall derive a relation where the corrective terms have the form of differential coefficients.

We begin by defining, by means of Bernoulli's polynomial $B_\nu(x)$, a *periodic* function $\bar{B}_\nu(x)$ with the period 1 which, in the interval $0 \leq x < 1$, is identical with $B_\nu(x)$. We call this periodic function

[1] Bromwich: The Theory of Infinite Series, p. 233.

the *periodic Bernoullian function of order* ν. It is completely determined for all real x, by having to satisfy the conditions

$$\left.\begin{aligned} \bar{B}_\nu(x) &= B_\nu(x) \text{ for } 0 \leq x < 1, \\ \bar{B}_\nu(x+1) &= \bar{B}_\nu(x) \text{ for all } x. \end{aligned}\right\} \quad (1)$$

If ν is different from 1, $\bar{B}_\nu(x)$ is evidently continuous for all x, as in that case $B_\nu(1) = B_\nu(0)$. On the other hand, $\bar{B}_1(x)$ has the appearance indicated in figure 3 and is, therefore, discontinuous at the points $0, \pm 1, \pm 2, \ldots$.

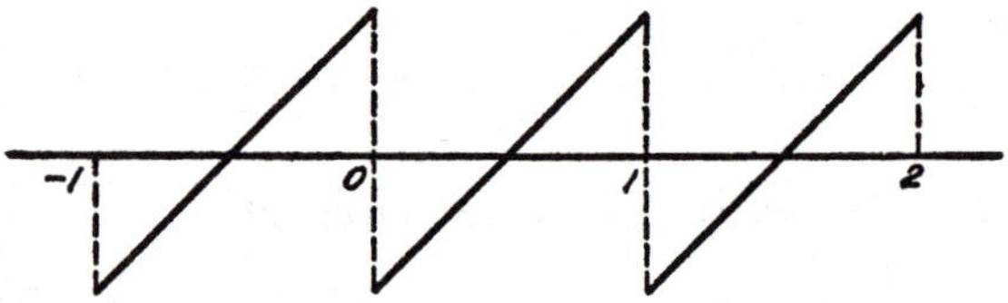

FIG. 3

From these properties of $\bar{B}_\nu(x)$ we may conclude that

$$D\bar{B}_\nu(x) = \nu \bar{B}_{\nu-1}(x) \qquad (\nu > 1), \tag{2}$$

so that $\bar{B}_\nu(x)$ possesses *continuous* differential coefficients of all the orders $1, 2, \ldots, \nu - 2$, while the differential coefficient of order $\nu - 1$ is discontinuous at the points $x = 0, \pm 1, \pm 2, \ldots$

145. We now consider the integral

$$R_m = -\int_0^1 \frac{\bar{B}_m(\theta - t)}{m!} f^{(m)}(x+t)\,dt. \tag{3}$$

We obtain from this, by partial integration, assuming $m > 1$,

$$R_m = -\frac{\bar{B}_m(\theta)}{m!}\left[f^{(m-1)}(x+1) - f^{(m-1)}(x)\right] + R_{m-1}$$

and, treating R_{m-1} in the same way, and so on,

$$R_m = -\sum_2^m \frac{\bar{B}_\nu(\theta)}{\nu!} \Delta f^{(\nu-1)}(x) + R_1.$$

We now assume

$$0 \leq \theta \leq 1, \tag{4}$$

so that instead of $\bar{B}_\nu(\theta)$ we may write $B_\nu(\theta)$. As $\bar{B}_1(\theta - t)$ is discontinuous for $t = \theta$, we write R_1 in the form

$$R_1 = -\int_0^\theta \bar{B}_1(\theta - t) f'(x + t)dt - \int_\theta^1 \bar{B}_1(\theta - t + 1) f'(x + t)dt,$$

whence, as $\bar{B}_1(x) = x - \frac{1}{2}$ for $0 \leq x < 1$,

$$R_1 = \int_0^1 tf'(x + t)dt - (\theta - \tfrac{1}{2}) \int_0^1 f'(x + t)dt - \int_\theta^1 f'(x + t)dt$$

$$= f(x + \theta) - B_1(\theta) \bigtriangleup f(x) - \int_x^{x+1} f(t)dt,$$

and consequently, inserting this expression in the above expression for R_m,

$$f(x + \theta) = \int_x^{x+1} f(t)dt + \sum_1^m \frac{B_\nu(\theta)}{\nu!} \bigtriangleup f^{(\nu - 1)}(x) + R_m. \tag{5}$$

This formula which is also valid for $m = 1$, is called *the general Euler-Maclaurin formula.*[1] We proved it in §13(17) for the case where $f(x)$ is a polynomial $P'(x)$, without having, in the special case, to put the restriction (4) on θ.

146. From (5) we may now easily derive the desired summation-formula. For, summing from $x = 0$ to $x = k - 1$ we obtain

$$\sum_{s=0}^{k-1} f(s + \theta) = \int_0^k f(t)\, dt + \sum_1^m \frac{B_\nu(\theta)}{\nu!} \Big[f^{(\nu - 1)}(x) \Big]_0^k + R, \tag{6}$$

where, according to (3),

$$R = -\sum_{s=0}^{k-1} \int_0^1 \frac{\bar{B}_m(\theta - t)}{m!} f^{(m)}(s + t)dt$$

$$= -\sum_{s=0}^{k-1} \int_s^{s+1} \frac{\bar{B}_m(\theta - t)}{m!} f^{(m)}(t)dt$$

or

$$R = -\int_0^k \frac{\bar{B}_m(\theta - t)}{m!} f^{(m)}(t)dt. \tag{7}$$

[1] See, for further particulars, N. E. Nörlund: Mémoire sur le calcul aux différences finies, Acta Mathematica, vol. 44, p. 98.

Formula (6) with the remainder-term (7) is called *the general Eulerian summation-formula.*

147. It is easy to indicate sufficient conditions for the validity of this formula for $k \to \infty$: If

$$1^0. \quad \sum_{s=0}^{\infty} f(s+\theta) \text{ and } \int_0^\infty f(t)dt \text{ converge,}$$

$$2^0. \ f^{(\nu-1)}(k) \to 0 \text{ for } k \to \infty \ (\nu = 1, 2, \ldots m),$$

we obtain from (6) and (7)

$$\left.\begin{aligned} \sum_{s=0}^{\infty} f(s+\theta) &= \int_0^\infty f(t)dt - \sum_{1}^{m} \frac{B_\nu(\theta)}{\nu!} f^{(\nu-1)}(0) + R, \\ R &= -\int_0^\infty \frac{\overline{B}_m(\theta - t)}{m\,!} f^{(m)}(t)dt. \end{aligned}\right\} \quad (8)$$

The convergence of the remainder-term follows from the conditions 1° and 2°.

148. The two most important particular cases of (6) are obtained for $\theta = 0$ and $\theta = \frac{1}{2}$. The practical importance of these cases is due to the fact that for these values of θ every other term of the expansion vanishes.

Assuming first $\theta = 0$, and putting $m = 2r$, the remainder-term (7) may, as $\overline{B}_{2r}(-t) = \overline{B}_{2r}(t)$, be written in the form

$$R = -\int_0^k \frac{\Delta_t \overline{B}_{2r}(0)}{(2r)\,!} f^{(2r)}(t)dt - \frac{B_{2r}}{(2r)\,!}\Big[f^{(2r-1)}(t)\Big]_0^k,$$

so that we obtain from (6)

$$\sum_{s=0}^{k-1} f(s) = \int_0^k f(t)dt - \tfrac{1}{2}\Big[f(x)\Big]_0^k + \sum_{\nu=1}^{r-1} \frac{B_{2\nu}}{(2\nu)\,!}\Big[f^{(2\nu-1)}(x)\Big]_0^k + R, \quad (9)$$

$$R = -\int_0^k \frac{\Delta_t \overline{B}_{2r}(0)}{(2r)\,!} f^{(2r)}(t)dt. \quad (10)$$

As $\triangle_t \overline{B}_{2r}(0)$ does not change its sign, we may apply the Theorem of Mean Value and find

$$R = -f^{(2r)}(\xi) \int_0^k \frac{\triangle_t \overline{B}_{2r}(0)}{(2r)!} dt$$

$$= -kf^{(2r)}(\xi) \int_0^1 \frac{\triangle_t B_{2r}(0)}{(2r)!} dt$$

or, as $\int_0^1 B_{2r}(t)dt = 0$,

$$R = \frac{kB_{2r}}{(2r)!} f^{(2r)}(\xi). \tag{11}$$

149. The Error-Test therefore leads to the simple result, that if $f^{(2r)}$ and $f^{(2r+2)}$ both keep their signs and have the same sign, then the error is numerically less than the first neglected term and has the same sign, assuming that this term does not vanish.

If it is only known that $f^{(2r)}$ does not change its sign, we may also establish a simple practical rule. As the numerically largest value of $\triangle_t \overline{B}_{2r}(0)$ is $\overline{B}_{2r}(\frac{1}{2}) - B_{2r} = (2^{1-2r} - 2)B_{2r}$, we find from (10), if θ is some number comprised between 0 and 1,

$$R = \theta(2 - 2^{1-2r}) \frac{B_{2r}}{(2r)!} \left[f^{(2r-1)}(t) \right]_0^k \qquad (0 < \theta < 1). \tag{12}$$

In particular, we have the convenient rule, that if $f^{(2r)}$ does not change its sign, the error is numerically less than *twice* the first neglected term, and has the same sign.

Both this rule and the rule derived from the Error-Test are applicable for $k \to \infty$.

150. Next, we put in (6), $\theta = \frac{1}{2}$, assuming again $m = 2r$. The remainder-term (7) may, as $\overline{B}_{2r}(\frac{1}{2} - t) = \overline{B}_{2r}(\frac{1}{2} + t)$, be written in the form

$$R = -\int_0^k \frac{\triangle_t \overline{B}_{2r}(\frac{1}{2})}{(2r)!} f^{(2r)}(t)dt - \frac{B_{2r}(\frac{1}{2})}{(2r)!} \left[f^{(2r-1)}(t) \right]_0^k,$$

so that we obtain, from (6),

$$\sum_{s=0}^{k-1} f(s+\tfrac{1}{2}) = \int_0^k f(t)dt + \sum_{\nu=1}^{r-1} \frac{B_{2\nu}(\frac{1}{2})}{(2\nu)!}\Big[f^{(2\nu-1)}(x)\Big]_0^k + R, \tag{13}$$

$$R = -\int_0^k \frac{\Delta_t \overline{B}_{2r}(\frac{1}{2})}{(2r)!} f^{(2r)}(t)dt. \tag{14}$$

As the numerically largest value which $B_{2r}(x) - B_{2r}$ can assume, is $B_{2r}(\frac{1}{2}) - B_{2r}$, we see, on writing the numerator in (14) in the form

$$[\overline{B}_{2r}(t+\tfrac{1}{2}) - B_{2r}] - [B_{2r}(\tfrac{1}{2}) - B_{2r}],$$

that it does not change its sign. We may therefore apply the Theorem of Mean Value and get

$$R = -f^{(2r)}(\xi)\int_0^k \frac{\Delta_t \overline{B}_{2r}(\frac{1}{2})}{(2r)!}dt$$

$$= -kf^{(2r)}(\xi)\int_0^1 \frac{\Delta_t \overline{B}_{2r}(\frac{1}{2})}{(2r)!}dt.$$

Now

$$\int_0^1 \overline{B}_{2r}(t+\tfrac{1}{2})\,dt = \frac{1}{2r+1}\Big[\overline{B}_{2r+1}(\tfrac{3}{2}) - \overline{B}_{2r+1}(\tfrac{1}{2})\Big] = 0,$$

consequently

$$R = \frac{kB_{2r}(\frac{1}{2})}{(2r)!} f^{(2r)}(\xi). \tag{15}$$

151. The Error-Test leads to the same rule as in the case of (11), as $B_{2r}(\frac{1}{2})$ and $B_{2r+2}(\frac{1}{2})$ have opposite signs.

If it is only known that $f^{(2r)}$ keeps its sign, we may here also establish a practical rule. For in this case we derive from (14), the numerator assuming its numerically largest value for $t = \frac{1}{2}$, the same expression as (12), only with the opposite sign, and consequently, introducing $B_{2r}(\frac{1}{2})$ instead of B_{2r},

$$R = \theta\left(2 + \frac{1}{2^{2r-1}-1}\right)\frac{B_{2r}(\frac{1}{2})}{(2r)!}\Big[f^{(2r-1)}(t)\Big]_0^k. \tag{16}$$

If $f^{(2r)}$ does not change its sign, the error is, therefore, numerically less than *three times* the first neglected term and has the same sign.

The convenient practical rules derived from (12) and (16) would seem to indicate that (9) is a more accurate formula than (13). As a matter of fact it is rather the other way round, as appears from a comparison of (11) and (15), although the difference is not great.

152. We shall refer to (9) and (13) as the *first* and the *second Eulerian summation-formula* respectively. They are so often used with the general limits of integration a and b instead of 0 and k, that we find it practical to state the results explicitly.

If, in (9)–(11), we put

$$f(x) = F(a + hx) \qquad \left(h = \frac{b-a}{k}\right), \tag{17}$$

we find

$$\left.\begin{aligned} h\sum_{s=0}^{k-1} F(a+sh) &= \int_a^b F(x)dx - \frac{h}{2}\Big[F(x)\Big]_a^b + \sum_{\nu=1}^{r-1}\frac{B_{2\nu}h^{2\nu}}{(2\nu)!}\Big[F^{(2\nu-1)}(x)\Big]_a^b + R, \\ R &= -h^{2r+1}\int_0^k \frac{\Delta_t \overline{B}_{2r}(0)}{(2r)!}F^{(2r)}(a+ht)\,dt \\ &= \frac{kB_{2r}h^{2r+1}}{(2r)!}F^{(2r)}(\xi). \end{aligned}\right\} \tag{18}$$

If, in (13)–(15), we make the same substitution, we obtain

$$\left.\begin{aligned} h\sum_{s=0}^{k-1} F\left(a+\frac{h}{2}+sh\right) &= \int_a^b F(x)dx + \sum_{\nu=1}^{r-1}\frac{B_{2\nu}(\frac{1}{2})h^{2\nu}}{(2\nu)!}\Big[F^{(2\nu-1)}(x)\Big]_a^b + R, \\ R &= -h^{2r+1}\int_0^k \frac{\Delta_t \overline{B}_{2r}(\frac{1}{2})}{(2r)!}F^{(2r)}(a+ht)dt \\ &= \frac{kB_{2r}(\frac{1}{2})h^{2r+1}}{(2r)!}F^{(2r)}(\xi). \end{aligned}\right\} \tag{19}$$

153. As a numerical example, let us calculate the same sum as in No. 127, making in (9) $k \to \infty$, and putting $f(x) = \dfrac{1}{(16 + x)^2}$. We thus obtain

$$\sum_{\nu = 16}^{\infty} \frac{1}{\nu^2} = \frac{1}{16} + \frac{1}{2}\frac{1}{16^2} + \frac{B_2}{16^3} + \frac{B_4}{16^5} + \frac{B_6}{16^7} + \ldots, \tag{20}$$

the error, stopping at a certain term, being smaller than the first neglected term, as $f^{(2r)}$ and $f^{(2r+2)}$ keep their signs and have the same sign. As already $\dfrac{B_4}{16^5}$ is numerically smaller than 4 units of the 8th decimal, we may, considering the degree of accuracy required, stop at B_2, and thus obtain $\sum\limits_{16}^{\infty} \dfrac{1}{\nu^2} = 0.0644938$, all the decimals being reliable. If, to this, we add the directly calculated sum $\sum\limits_{1}^{15} \dfrac{1}{\nu^2} = 1.5804403$, we get $\sum\limits_{1}^{\infty} \dfrac{1}{\nu^2} = 1.6449341$, the first 6 decimals being reliable, the last one, as a matter of fact, being also correct.

It follows from §13 (38) that the expansion employed for the calculation of $\sum\limits_{16}^{\infty} \dfrac{1}{\nu^2}$ is divergent if continued indefinitely; which, as we have seen, does not prevent its being used for numerical calculations, provided we can indicate limits to the error involved. We may even in this way calculate $\sum\limits_{1}^{\infty} \dfrac{1}{\nu^2}$ to any given number of decimals, if we begin by calculating a suitable number of terms directly, and thereafter apply the summation-formula.

154. Eulerian formulas for repeated summation or integration are derived in the same way as the corresponding Gaussian formulas, see No. 130. We content ourselves with deriving a special formula which has applications to actuarial problems.

We find from §11(30) for $k = 1$, in the case of convergence,

$$\int_x^{\infty} \int_x^{\infty} F(x) dx^2 = \int_x^{\infty} (t - x) F(t) dt. \tag{21}$$

If, in (8), we put $\theta = 0$ and replace $f(\tau)$ by $f(\tau + x)$, we get

$$\int_x^\infty f(t)dt = \sum_x^\infty f(\nu) + \sum_{\nu=1}^{m} \frac{B_\nu}{\nu\,!} f^{(\nu-1)}(x) + R,$$

$$R = \int_0^\infty \frac{\bar{B}_m(-t)}{m\,!} f^{(m)}(t+x)dt.$$

In this formula, we put

$$f(t) = (t-x)\,F(t),$$

so that

$$f^{(n)}(t) = (t-x)F^{(n)}(t) + n\,F^{(n-1)}(t),$$

and

$$f^{(\nu-1)}(x) = (\nu-1)\,F^{(\nu-2)}(x),$$

and obtain by (21)

$$\int^\infty \int^\infty F(x)dx^2 = \sum_{\nu=x+1}^{\infty} (\nu - x)\,F(\nu) + \sum_{\nu=2}^{m} \frac{B_\nu}{\nu\,!} (\nu - 1)\,F^{(\nu-2)}(x) + R$$

or, by §11(14),

$$\int_x^\infty \int_x^\infty F(x)dx^2 = \Sigma''\,F(x-1) + \sum_{\nu=2}^{m} \frac{B_\nu}{\nu} \frac{F^{(\nu-2)}(x)}{(\nu-2)\,!} + R,$$

$$R = \int_0^\infty \frac{\bar{B}_m(-t)}{m\,!} [t\,F^{(m)}(t+x) + m\,F^{(m-1)}(t+x)]\,dt.$$

We now put $m = 2r$ and

$$\bar{B}_{2r}(t) = \triangle_t \bar{B}_{2r}(0) + B_{2r}.$$

Hence we have, for the remainder-term,

$$\left.\begin{aligned} R = \int_0^\infty \frac{\triangle_t \bar{B}_{2r}(0)}{(2r)\,!} [t\,F^{(2r)}(t+x) + 2r\,F^{(2r-1)}(t+x)]dt \\ + \frac{B_{2r}}{(2r)\,!} \int_0^\infty [t\,F^{(2r)}(t+x) + 2r\,F^{(2r-1)}(t+x)]\,dt; \end{aligned}\right\}$$

but the second integration on the right can be carried out, the result being

$$\int_0^\infty [t\,F^{(2r)}(t+x) + 2r\,F^{(2r-1)}(t+x)]dt = -\,(2r-1)\,F^{(2r-2)}(x).$$

Inserting this in the expression for the remainder-term, we obtain the final formula

$$\int_x^\infty \int_x^\infty F(x)dx^2 = \Sigma''\,F(x-1) + \sum_{\nu=1}^{r-1} \frac{B_{2\nu}}{2\nu}\,\frac{F^{(2\nu-2)}(x)}{(2\nu-2)\,!} + R, \quad (22)$$

$$R = \int_0^\infty \frac{\Delta_t \overline{B}_{2r}(0)}{(2r)\,!}\,[t\,F^{(2r)}(t+x) + 2r\,F^{(2r-1)}(t+x)]dt. \quad (23)$$

This formula is, according to No. 147, valid, provided that:

1. $\int_x^\infty \int_x^\infty F(x)dx^2$ and $\Sigma''\,F(x-1)$ converge;

2. $t\,F^{(n)}\,(t) \to 0$ for $t \to \infty$ $(n = 0, 1, 2, \ldots, 2r-1)$.

The convergence of the remainder-term follows from 1 and 2.

§15. Lubbock's and Woolhouse's Formulas

155. If a sum, containing a great many terms, is to be calculated, it is natural, as a first approximation to the required sum to take, say, h times the sum of every h^{th} term. To this must, then, be added certain corrective terms; and these may either be expressed as finite differences, or as differential coefficients. In the former case, we have formulas of Lubbock's type; in the latter, formulas of Woolhouse's type. The summation-formulas of Laplace's and Euler's types may evidently be regarded as limiting cases of the types dealt with in this section; we have, however, found it more practical to derive them separately.

156. We obtain from the interpolation-formula with descending differences

$$f\left(n + \frac{s}{h}\right) = \sum_{\nu=0}^{r-1} \frac{1}{\nu!}\left(\frac{s}{h}\right)^{(\nu)} \Delta^\nu f(n) + \frac{1}{r!}\left(\frac{s}{h}\right)^{(r)} f^{(r)}(\xi). \quad (1)$$

We now write, for abbreviation,

$$A_\nu = \frac{1}{\nu!} \sum_{s=0}^{h-1} \left(\frac{s}{h}\right)^{(\nu)} \tag{2}$$

and sum (1) from $s = 0$ to $s = h - 1$. As $\left(\frac{s}{h}\right)^{(r)}$ does not change its sign for the values of s employed, we may apply the Theorem of Mean Value to the remainder-term and find

$$\sum_{s=0}^{h-1} f\left(n + \frac{s}{h}\right) = \sum_{\nu=0}^{r-1} A_\nu \Delta^\nu f(n) + A_r f^{(r)}(\xi).$$

We see immediately from (2) that $A_0 = h$. Keeping the first term on the right apart, and summing from $n = 0$ to $n = k - 1$, we therefore obtain

$$\sum_{\nu=0}^{kh-1} f\left(\frac{\nu}{h}\right) = h \sum_{0}^{k-1} f(\nu) + \sum_{\nu=1}^{r-1} A_\nu \left[\Delta^{\nu-1} f(x)\right]_0^k + kA_r f^{(r)}(\xi). \tag{3}$$

In this formula, the larger interval has been taken as unity. In practice, it is as a rule more convenient to choose the smaller interval as unity which may be done by putting $f(x) = F(hx)$. Introducing the notation

$$\Delta_h = E^h - 1,$$

the resulting formula may be written

$$\sum_{0}^{kh-1} F(\nu) = h \sum_{\nu=0}^{k-1} F(h\nu) + \sum_{\nu=1}^{r-1} A_\nu \left[\Delta_h^{\nu-1} F(x)\right]_0^{kh} + kh^r A_r F^{(r)}(\xi). \tag{4}$$

This is Lubbock's formula with its remainder-term.

We still have to calculate the coefficients A_ν. This may be done in several ways, e.g. by expanding $\left(\frac{s}{h}\right)^{(\nu)}$ in powers of s and summing by Bernoulli's polynomials. But we may also find a recurrence formula for the A_ν. An examination of (2) shows that A_ν is the coefficient of x^ν in the expansion of

$$\sum_{s=0}^{h-1} (1 + x)^{\frac{s}{h}}$$

which is easily summed, so that

$$\frac{x}{(1 + x)^{\frac{1}{h}} - 1} = \Lambda_0 + x\Lambda_1 + x^2\Lambda_2 + \ldots , \qquad (5)$$

whence

$$x = (\Lambda_0 + x\Lambda_1 + x^2\Lambda_2 + \ldots) \left[\binom{\frac{1}{h}}{1} x + \binom{\frac{1}{h}}{2} x^2 + \binom{\frac{1}{h}}{3} x^3 + \ldots \right].$$

As the coefficient of x^ν on the right vanishes for $\nu > 1$, we have in addition to $\Lambda_0 = h$,

$$\binom{\frac{1}{h}}{1} \Lambda_\nu + \binom{\frac{1}{h}}{2} \Lambda_{\nu - 1} + \binom{\frac{1}{h}}{3} \Lambda_{\nu - 2} + \ldots + \binom{\frac{1}{h}}{\nu + 1} \Lambda_0 = 0 \qquad (6)$$

which serves for the successive calculation of the coefficients Λ_ν.
We find

$$\Lambda_0 = h \qquad\qquad \Lambda_4 = -\frac{(h^2 - 1)(19\,h^2 - 1)}{720\,h^3}$$

$$\Lambda_1 = \frac{h - 1}{2} \qquad\qquad \Lambda_5 = \frac{(h^2 - 1)(9\,h^2 - 1)}{480\,h^3}$$

$$\Lambda_2 = -\frac{h^2 - 1}{12\,h} \qquad\qquad \Lambda_6 = -\frac{(h^2 - 1)(863\,h^4 - 145\,h^2 + 2)}{60480\,h^5}$$

$$\Lambda_3 = \frac{h^2 - 1}{24\,h} \qquad\qquad \Lambda_7 = \frac{(h^2 - 1)(275\,h^4 - 61\,h^2 + 2)}{24192\,h^5}$$

We give below a table of the numerical values of these expressions as far as $h = 11$.

h	$-\Lambda_2$		Λ_3		$-\Lambda_4$	
2	0.125		0.0625		0.03906	25
3	0.22222	22222	0.11111	11111	0.06995	88477
4	0.3125		0.15625		0.09863	28125
5	0.4		0.2		0.1264	
6	0.48611	11111	0.24305	55556	0.15371	01337
7	0.57142	85714	0.28571	42857	0.18075	80175
8	0.65625		0.32812	5	0.20764	16016
9	0.74074	07407	0.37037	03704	0.23441	54854
10	0.825		0.4125		0.26111	25
11	0.90909	09091	0.45454	54545	0.28775	35687

h	Λ_5		$-\Lambda_6$		Λ_7	
2	0.02734	375	0.02050	78125	0.01611	32812
3	0.04938	27160	0.03734	18686	0.02956	86633
4	0.06982	42188	0.05294	79980	0.04203	79639
5	0.0896		0.06803	2	0.05408	
6	0.10903	74228	0.08284	85499	0.06590	15960
7	0.12827	98834	0.09751	03911	0.07759	52197
8	0.14739	99023	0.11207	48520	0.08920	81261
9	0.16643	80430	0.12657	41007	0.10076	69086
10	0.18541	875	0.14102	75625	0.11228	76562
11	0.20435	76258	0.15544	76532	0.12378	06658

157. It is seen from (6), that Λ_ν is a rational function of h. It is easy to find limits between which Λ_ν is comprised. Let us assume $\nu > 2$ and apply the Theorem of Mean Value to (2), retaining under the sign Σ the factor $\frac{s}{h}\left(\frac{s}{h} - 1\right)$. As

$$\sum_{s=0}^{h-1} \frac{s}{h}\left(\frac{s}{h} - 1\right) = 2\Lambda_2,$$

we get

$$\Lambda_\nu = \frac{(-1)^\nu}{\nu!}\left(2 - \frac{\xi}{h}\right)\left(3 - \frac{\xi}{h}\right)\cdots\left(\nu - 1 - \frac{\xi}{h}\right) 2\Lambda_2, \tag{7}$$

ξ being comprised between 0 and $h - 1$.

As Λ_2 is negative, it is seen that Λ_ν (apart from Λ_0) has the sign $(-1)^{\nu+1}$.

From (7) we obtain simple limits to Λ_ν by putting $\xi = 0$ and $\xi = h$, viz.

$$\frac{-2\Lambda_2}{\nu(\nu-1)} < |\Lambda_\nu| < \frac{-2\Lambda_2}{\nu} \qquad (\nu > 2). \tag{8}$$

158. By Stirling's and Bessel's interpolation formulas, we obtain certain formulas which are more suited for numerical calculations than Lubbock's original formula.

We have by Stirling's formula

$$\left.\begin{aligned} f\left(n+\frac{x}{h}\right) &= \sum_{\nu=0}^{r-1} \frac{1}{(2\nu)!}\left(\frac{x}{h}\right)^{[2\nu]} \delta^{2\nu} f(n) \\ &\quad + \sum_{\nu=1}^{r-1} \frac{1}{(2\nu-1)!}\left(\frac{x}{h}\right)^{[2\nu]-1} \square\, \delta^{2\nu-1} f(n) \\ &\quad + \left(\frac{x}{h}\right)^{[2r]-1} f\left(n+\frac{x}{h}, n, n \pm 1, \ldots, n \pm (r-1)\right). \end{aligned}\right\}$$

We now assume that h is a positive integer, and put

$$P_{2\nu} = \frac{1}{(2\nu)!} \sum_{-(h-1)/2}^{(h-1)/2} \left(\frac{x}{h}\right)^{[2\nu]}, \tag{9}$$

the summation referring to x, and x assuming in succession the values

$$-(h-1)/2, 1-(h-1)/2, 2-(h-1)/2, \ldots, (h-1)/2.$$

Summing $f\left(n+\frac{x}{h}\right)$ with regard to these values, we get

$$\sum_{-(h-1)/2}^{(h-1)/2} f\left(n+\frac{x}{h}\right) = \sum_{\nu=0}^{r-1} P_{2\nu}\, \delta^{2\nu} f(n) + R,$$

$$R = \sum_{-(h-1)/2}^{(h-1)/2} \left(\frac{x}{h}\right)^{[2r]-1} f\left(n+\frac{x}{h}, n, n \pm 1, \ldots, n \pm (r-1)\right).$$

But if, in this expression, we write $-x$ instead of x, we see that R may also be written in the form

$$R = -\sum_{-(h-1)/2}^{(h-1)/2} \left(\frac{x}{h}\right)^{[2r]-1} f\left(n-\frac{x}{h}, n, n \pm 1, \ldots, n \pm (r-1)\right),$$

and forming the arithmetical mean of these two expressions, we find, exactly as in No. 30,

$$R = \frac{1}{(2r)!} \sum_{-(h-1)/2}^{(h-1)/2} \left(\frac{x}{h}\right)^{[2r]} f^{(2r)}(\xi)$$

or, as $\left(\frac{x}{h}\right)^{[2r]}$ does not change its sign for the values of x employed, we have by the Theorem of Mean Value,

$$R = P_{2r} f^{(2r)}(\xi).$$

We have, therefore, proved the formula

$$\sum_{-(h-1)/2}^{(h-1)/2} f\left(n + \frac{x}{h}\right) = \sum_{\nu=0}^{r-1} P_{2\nu}\, \delta^{2\nu} f(n) + P_{2r} f^{(2r)}(\xi).$$

But from this we obtain by summation with regard to n from $n = 0$ to $n = k - 1$, keeping the term containing $P_0\,(= h)$ apart,

$$\left.\begin{aligned} \sum_{\nu=-(h-1)/2}^{hk-(h+1)/2} f\left(\frac{\nu}{h}\right) &= h \sum_{0}^{k-1} f(\nu) + \sum_{\nu=1}^{r-1} P_{2\nu} \Big[\delta^{2\nu-1} f(x)\Big]_{-\frac{1}{2}}^{k-\frac{1}{2}} + R \\ R &= k P_{2r} f^{(2r)}(\xi). \end{aligned}\right\}(10)$$

If, finally, we put

$$f(x) = F\left(h x + \frac{h-1}{2}\right)$$

and

$$\delta_h = E^{\frac{h}{2}} - E^{-\frac{h}{2}},$$

we obtain the more practical form

$$\left.\begin{aligned} \sum_{0}^{hk-1} F(\nu) &= h \sum_{\nu=0}^{k-1} F\left(h\nu + \frac{h-1}{2}\right) + \sum_{\nu=1}^{r-1} P_{2\nu} \Big[\delta_h^{2\nu-1} F(x)\Big]_{-\frac{1}{2}}^{hk-\frac{1}{2}} + R \\ R &= k P_{2r} h^{2r} F^{(2r)}(\xi). \end{aligned}\right\}(11)$$

We have still to calculate the coefficients $P_{2\nu}$. We find from (9), developing in powers of x, for $\nu > 0$

$$P_{2\nu} = \frac{1}{(2\nu)!} \sum_{-(h-1)/2}^{(h-1)/2} \sum_{\mu=1}^{\nu} \left(\frac{x}{h}\right)^{2\mu} \frac{D^{2\mu}\, 0^{[2\nu]}}{(2\mu)!}$$

$$= \frac{1}{(2\nu)!} \sum_{\mu=1}^{\nu} \frac{D^{2\mu}\, 0^{[2\nu]}}{h^{2\mu}\,(2\mu)!} \sum_{-(h-1)/2}^{(h-1)/2} x^{2\mu};$$

but by § 13 (21) we have for $x = -\dfrac{h-1}{2}$, $m = 2\mu$, $n = h$

$$\sum_{-(h-1)/2}^{(h-1)/2} x^{2\mu}$$

$$= \int_0^h B_{2\mu}\left(t - \frac{h}{2} + \frac{1}{2}\right)dt$$

$$= \int_{-\frac{h}{2}}^{\frac{h}{2}} B_{2\mu}\left(t + \frac{1}{2}\right)dt$$

$$= 2\int_0^{\frac{h}{2}} B_{2\mu}\left(t + \frac{1}{2}\right)dt$$

$$= \frac{2}{2\mu+1} B_{2\mu+1}\left(\frac{h+1}{2}\right),$$

so that, for $\nu > 0$

$$P_{2\nu} = \frac{2}{(2\nu)!} \sum_{\mu=1}^{\nu} \frac{D^{2\mu}0^{[2\nu]}}{h^{2\mu}\,(2\mu+1)!} B_{2\mu+1}\left(\frac{h+1}{2}\right). \tag{12}$$

These coefficients may also be calculated by the recurrence formula §18 (48). It is seen immediately from (9) that $P_{2\nu}$ (apart from P_0) has the sign $(-1)^{\nu+1}$. We find for the first few values:

$$P_0 = h;\ P_2 = \frac{h^2-1}{24\,h};\ P_4 = -\frac{(h^2-1)\,(17\,h^2+7)}{5760\,h^3};$$

$$P_6 = \frac{(h^2-1)\,(367\,h^4+178\,h^2+31)}{967680\,h^5}.$$

A table of the numerical values of these coefficients for various values of h is given below.

h	P_2	$-P_4$	P_6
2	0.0625	0.00488 28125	0.00064 08691
3	0.11111 11111	0.00823 04527	0.00106 69105
4	0.15625	0.01135 25391	0.00146 57974
5	0.2	0.0144	0.00185 6
6	0.24305 55556	0.01741 33552	0.00224 22861
7	0.28571 42857	0.02040 81633	0.00262 64567
8	0.32812 5	0.02339 17236	0.00300 93491
9	0.37037 03704	0.02636 79317	0.00339 14081
10	0.4125	0.02933 90625	0.00377 28926
11	0.45454 54545	0.03230 65364	0.00415 39637

We shall refer to the relation (11) as *the second formula of Lubbock's type.* In comparison with (4), it offers the advantage that the corrective terms only contain differences of odd order. In practice, it is as a rule only applicable if h is an odd integer, $h = 2m + 1$; as for $h = 2m$ use is made of the values of $F(2m\nu + m - \frac{1}{2})$ which are not found in the given table.

159. A similar formula may be found by Bessel's formula as follows. We have by this formula

$$
\begin{aligned}
\left(n+\tfrac{1}{2}+\frac{x}{h}\right) = & \sum_{\nu=0}^{r-1} \frac{1}{(2\nu)!}\left(\frac{x}{h}\right)^{[2\nu+1]-1} \square\, \delta^{2\nu} f(n+\tfrac{1}{2}) \\
& + \sum_{\nu=0}^{r-1} \frac{1}{(2\nu+1)!}\left(\frac{x}{h}\right)^{[2\nu+1]} \delta^{2\nu+1} f(n+\tfrac{1}{2}) \\
& + \left(\frac{x}{h}\right)^{[2r+1]-1} f\left(n+\tfrac{1}{2}+\frac{x}{h}, n, n\pm 1, \ldots, n\pm(r-1), n+r\right).
\end{aligned}
$$

Assuming h positive and integral, and putting

$$
Q_{2\nu} = \frac{1}{(2\nu)!} \sum_{-(h-2)/2}^{(h-2)/2} \left(\frac{x}{h}\right)^{[2\nu+1]-1}, \tag{13}
$$

we find

$$\sum_{-(h-2)/2}^{(h-2)/2} f\left(n+\tfrac{1}{2}+\frac{x}{h}\right) = \sum_{\nu=0}^{r-1} Q_{2\nu} \square\, \delta^{2\nu} f(n+\tfrac{1}{2}) + R,$$

$$R = \sum_{-(h-2)/2}^{(h-2)/2} \binom{x}{h}^{[2r+1]-1} f\left(n+\tfrac{1}{2}+\frac{x}{h}, n, n\pm 1, \ldots, n\pm(r-1), n+r\right).$$

As $\binom{x}{h}^{[2r+1]-1}$ does not change its sign for the values of x employed, we may apply the Theorem of Mean Value and get

$$R = Q_{2r} f^{(2r)}(\xi).$$

Keeping the term containing $Q_0(= h-1)$ apart, the formula may be written

$$\left.\begin{aligned} \sum_{-h/2}^{h/2-1} f\left(n+\frac{1}{2}+\frac{x}{h}\right) &= \frac{h}{2} f(n) + \frac{h}{2} f(n+1) - \frac{1}{2} \triangle f(n) \\ &+ \sum_{\nu=1}^{r-1} Q_{2\nu} \square\, \delta^{2\nu} f\left(n+\frac{1}{2}\right) + Q_{2r} f^{(2r)}(\xi). \end{aligned}\right\}$$

Summing now with regard to n from $n = 0$ to $n = k-1$, we obtain

$$\left.\begin{aligned} \sum_{\nu=0}^{hk-1} f\left(\frac{\nu}{h}\right) &= h \sum_{0}^{k-1} f(\nu) + \frac{h-1}{2}\Big[f(x)\Big]_0^k + \sum_{\nu=1}^{r-1} Q_{2\nu}\Big[\square\, \delta^{2\nu-1} f(x)\Big]_0^k + R \\ R &= k\, Q_{2r} f^{(2r)}(\xi). \end{aligned}\right\} \quad (14)$$

If, finally, we put $f(x) = F(hx)$, we find the more practical form

$$\left.\begin{aligned} \sum_{0}^{hk-1} F(\nu) &= h \sum_{\nu=0}^{k-1} F(h\nu) + \frac{h-1}{2}\Big[F(x)\Big]_0^{hk} + \sum_{\nu=1}^{r-1} Q_{2\nu}\Big[\square_h\, \delta_h^{2\nu-1} F(x)\Big]_0^{hk} + R \\ R &= k\, Q_{2r} h^{2r} F^{(2r)}(\xi). \end{aligned}\right\} \quad (15)$$

The coefficients $Q_{2\nu}$ may be calculated as follows. We find from (13), developing in powers of x,

$$Q_{2\nu} = \frac{1}{(2\nu)!} \sum_{-(h-2)/2}^{(h-2)/2} \sum_{\mu=0}^{\nu} \left(\frac{x}{h}\right)^{2\mu} \frac{D^{2\mu} 0^{[2\nu+1]-1}}{(2\mu)!};$$

but on comparing the expansions of $x^{[n]-1}$ and $x^{[n]}$ in powers of x, it is seen that

$$D^s 0^{[n]-1} = \frac{1}{s+1} D^{s+1} 0^{[n]},$$

so that

$$Q_{2\nu} = \frac{1}{(2\nu)!} \sum_{\mu=0}^{\nu} \frac{D^{2\mu+1} 0^{[2\nu+1]}}{h^{2\mu}(2\mu+1)!} \sum_{-(h-2)/2}^{(h-2)/2} x^{2\mu}.$$

We therefore have

$$Q_{2\nu} = \frac{2}{(2\nu)!} \sum_{\mu=0}^{\nu} \frac{D^{2\mu+1} 0^{[2\nu+1]}}{h^{2\mu}(2\mu+1)!(2\mu+1)} B_{2\mu+1}\left(\frac{h}{2}\right). \tag{16}$$

These coefficients may also be calculated by the recurrence formula §18 (51). It is seen from (13) that $Q_{2\nu}$ has the sign $(-1)^\nu$. The first few values are:

$$Q_0 = h - 1; Q_2 = -\frac{h^2 - 1}{12h}; Q_4 = \frac{(h^2-1)(11h^2+1)}{720\,h^3};$$

$$Q_6 = -\frac{(h^2-1)(191h^4 + 23h^2 + 2)}{60480h^5}.$$

A table of the numerical values of these coefficients is given below.

h	$-Q_2$		Q_4		$-Q_6$	
2	0.125		0.02343	75	0.00488	28125
3	0.22222	22222	0.04115	22634	0.00853	52843
4	0.3125		0.05761	71875	0.01193	23730
5	0.4		0.0736		0.01523	2
6	0.48611	11111	0.08934	54218	0.01848	38380
7	0.57142	85714	0.10495	62682	0.02170	86418
8	0.65625		0.12048	33984	0.02491	66489
9	0.74074	07407	0.13595	48849	0.02811	35002
10	0.825		0.15138	75	0.03130	25625
11	0.90909	09091	0.16679	18858	0.03448	59703

We shall refer to (15) as *the third formula of Lubbock's type.* It only contains differences of odd order and is, therefore, preferable to Lubbock's original formula. In comparison with (11), the second formula of Lubbock's type, the latter is as a rule to be preferred, as the coefficients decrease more rapidly, and the formation of arithmetical means of differences is avoided. But if h is an even integer, $h = 2m$, (11) is inconvenient while (15) may still be employed.[1]

160. Limits to the coefficients $P_{2\nu}$ and $Q_{2\nu}$ may be found by a similar process as that employed in the case of Λ_ν. Thus, we obtain from (9), assuming $\nu > 1$ and $0 < \theta < \frac{1}{4}$,

$$P_{2\nu} = \frac{(-1)^{\nu-1}}{(2\nu)!}(1^2 - \theta)(2^2 - \theta) \ldots [(\nu - 1)^2 - \theta]\, 2P_2$$

showing that these coefficients have alternating signs and (for $\theta = 0$) that

$$|P_{2\nu}| \leq \frac{2P_2}{\nu^2\binom{2\nu}{\nu}}.$$

Similarly, we obtain from (13), assuming $\nu > 1$ and $0 < \theta < \frac{1}{4}$,

$$Q_{2\nu} = (-1)^{\nu-1}\frac{2}{(2\nu)!}(\tfrac{9}{4} - \theta) \ldots \left[\frac{(2\nu - 1)^2}{4} - \theta\right] Q_2$$

showing that the signs are alternating and that

$$|Q_{2\nu}| \leq -Q_2\binom{2\nu}{\nu}2^{-(4\nu-3)}.$$

161. In order to deduce Woolhouse's formula we put, in §14 (9), $k = mn$, and in §14 (18), $a = 0$, $b = mn$, $h = n$, $k = m$, $F = f$, and subtract the two formulas from each other. We find, replacing s by ν,

$$\sum_{0}^{mn-1} f(\nu) = n\sum_{\nu=0}^{m-1} f(n\nu) + \frac{n-1}{2}\Big[f(x)\Big]_0^{mn} - \sum_{\nu=1}^{r-1}\frac{B_{2\nu}}{(2\nu)!}(n^{2\nu} - 1)\Big[f^{(2\nu-1)}(x)\Big]_0^{mn} + R, \quad (17)$$

$$R = n\int_0^m \frac{n^{2r}\Delta_t \overline{B}_{2r}(0) - \Delta_{nt}\overline{B}_{2r}(0)}{(2r)!} f^{(2r)}(nt)dt. \quad (18)$$

In order to be able to apply the Theorem of Mean Value to this integral, we proceed to show that the function

$$n^{2r}\Delta_t \overline{B}_{2r}(0) - \Delta_{nt}\overline{B}_{2r}(0) \quad (19)$$

[1] See the paper by I. Lehmann quoted in No. 22.

does not change its sign, as $n^{2r}\Delta_t\overline{B}_{2r}(0)$ is always numerically greater than, or equal to $\Delta_{nt}\overline{B}_{2r}(0)$, so that the sign is $(-1)^r$.

In proving this, we may confine ourselves to the interval $0 \le t \le \frac{1}{2n}$; for the function (19) is periodic with the period 1 and symmetrical about $t = \frac{1}{2}$; and for $0 \le t < \frac{1}{2}$ the function $n^{2r}\Delta_t\overline{B}_{2r}(0)$ is numerically increasing with t, while $\Delta_{nt}\overline{B}_{2r}(0)$ attains its numerically largest value for $t = \frac{1}{2n}$ (see fig. 4).

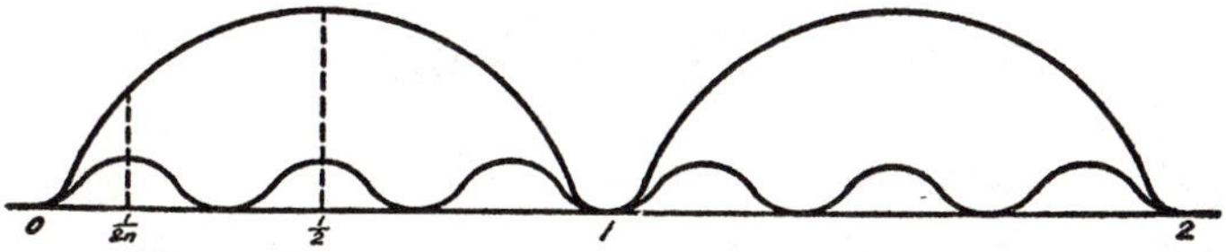

FIG. 4

But in the interval $0 \le t \le \frac{1}{n}$, (19) is identical with

$$n^{2r}\Delta_t B_{2r}(0) - \Delta_{nt}B_{2r}(0),$$

and according to §13 (28) this function, the coefficients $G_\nu^{(r)}$ being all positive, does not change its sign provided that the expression

$$n^{2r}(t - t^2)^k - (nt - n^2t^2)^k$$

keeps its sign. But this expression is positive for $0 < t < \frac{1}{n}$, as is seen by writing it in the form

$$n^k t^k [n^{2r-k}(1 - t)^k - (1 - nt)^k];$$

and the proof is complete.

We therefore obtain from (18)

$$R = nf^{(2r)}(\xi) \int_0^m \frac{n^{2r}\Delta_t\overline{B}_{2r}(0) - \Delta_{nt}\overline{B}_{2r}(0)}{(2r)!}\,dt.$$

In order to calculate the integral, we note that if $\omega(t)$ is a periodic function with the period 1, and n and m are positive integers, then

$$\int_0^m \omega(nt)dt = \frac{1}{n}\int_0^{mn} \omega(t)dt = m\int_0^1 \omega(t)dt.$$

Hence

$$R = \frac{nm}{(2r)!} f^{(2r)}(\xi)\,(n^{2r}-1)\int_0^1 \Delta_t B_{2r}(0)\,dt,$$

and consequently, by §13 (2),

$$R = -\frac{B_{2r}}{(2r)!} mn(n^{2r}-1) f^{(2r)}(\xi), \tag{20}$$

being the practical form of the remainder-term.

Woolhouse's formula (17) with the remainder-term (20) is preferable to the formulas of Lubbock's type in those cases where it is easy to calculate the differential coefficients of $f(x)$.

162. A variant of Woolhouse's formula is obtained if, in §14 (13), we put $k = mn$, and replace $f(\tau)$ by $f(\tau - \frac{1}{2})$; in §14 (19) put $a = -\frac{1}{2}$, $b = mn - \frac{1}{2}$, $h = n$, $k = m$, $F = f$; and subtract the two formulas from each other. We thus obtain, replacing s by ν,

$$\sum_0^{mn-1} f(\nu) = n\sum_{\nu=0}^{m-1} f\left(n\nu + \frac{n-1}{2}\right) - \sum_{\nu=1}^{r-1}\frac{B_{2\nu}(\frac{1}{2})}{(2\nu)!}(n^{2\nu}-1)\Big[f^{(2\nu-1)}(x)\Big]_{-\frac{1}{2}}^{mn-\frac{1}{2}} + R \tag{21}$$

$$R = n\int_0^m \frac{n^{2r}\Delta_t \bar{B}_{2r}(\frac{1}{2}) - \Delta_{nt}\bar{B}_{2r}(\frac{1}{2})}{(2r)!} f^{(2r)}(nt - \tfrac{1}{2})dt. \tag{22}$$

In order to be able to apply the Theorem of Mean Value we proceed to show that the function

$$n^{2r}\Delta_t \bar{B}_{2r}(\tfrac{1}{2}) - \Delta_{nt}\bar{B}_{2r}(\tfrac{1}{2}) \tag{23}$$

does not change its sign. For similar reasons as in the case of (19) we may confine ourselves to the interval $0 \le t \le \frac{1}{2n}$. In this interval (23) is identical with

$$n^{2r}\Delta_t B_{2r}(\tfrac{1}{2}) - \Delta_{nt} B_{2r}(\tfrac{1}{2}). \tag{24}$$

For $r = 1$ it is easily proved that this expression vanishes identically and therefore does not change its sign.

For $r > 1$ we note that the function (24) vanishes for $t = 0$, and that its differential coefficient, apart from the constant factor $2rn$, is

$$n^{2r-1}B_{2r-1}(t + \tfrac{1}{2}) - B_{2r-1}(nt + \tfrac{1}{2}). \tag{25}$$

We need only show that this expression does not change its sign in the interval.

Now we have by §13 (31) for $r > 1$

$$B_{2r-1}(x + \tfrac{1}{2}) = (-1)^{r+1}2x \sum_{\nu=0}^{r-2} H_\nu^{(r)}(\tfrac{1}{4} - x^2)^{r-\nu-1},$$

the coefficients $H_\nu^{(r)}$ being all positive. The function (25) will therefore keep its sign, if the expression

$$n^{2r-1}t(\tfrac{1}{4} - t^2)^k - nt(\tfrac{1}{4} - n^2t^2)^k$$

keeps its sign; but this expression is evidently positive in the interval.

Having thus proved that the function (23) does not change its sign, we obtain from (22)

$$R = nf^{(2r)}(\xi) \int_0^m \frac{n^{2r}\Delta_t\overline{B}_{2r}(\frac{1}{2}) - \Delta_{nt}\overline{B}_{2r}(\frac{1}{2})}{(2r)!}dt$$

$$= mnf^{(2r)}(\xi)\,(n^{2r} - 1)\int_0^1 \frac{\Delta_t\overline{B}_{2r}(\frac{1}{2})}{(2r)!}dt$$

or finally

$$R = -\frac{B_{2r}(\frac{1}{2})}{(2r)!}mn(n^{2r} - 1)f^{(2r)}(\xi). \tag{26}$$

We shall refer to (21) with its remainder-term as *the second formula of Woolhouse's type.*

It should be noted that (21) is only of real practical value, if n is an odd integer $n = 2k + 1$, as for $n = 2k$, terms of the form $f(2k\nu + k - \frac{1}{2})$ occur in the formula.

163. The application of the Error-Test to the formulas developed in this section, is immediate. Confining ourselves to the case where the first neglected term does not vanish, the conditions for the error to be numerically smaller than this term, and to have the same sign, are briefly the following:

$$\begin{aligned} &\textit{Lubbock I:} && f^{(r)} f^{(r+1)} > 0. \\ &\text{— II \& III:} && F^{(2r)} F^{(2r+2)} > 0. \\ &\textit{Woolhouse } \text{I \& II:} && f^{(2r)} f^{(2r+2)} > 0. \end{aligned}$$

164. Exactly as in the case of the summation-formulas of §12 and §14, we may let the upper limits of summation tend to infinity, provided that the sums converge and (in the case of the formulas of Woolhouse's type) the differential coefficients vanish at the upper limit, and the Error-Test remains applicable under these circumstances.

165. As a numerical example we propose to calculate $\sum\limits_{200}^{349} \frac{1}{\nu}$ by (11), putting $k = 10$, $h = 15$, $F(\nu) = \frac{1}{200 + \nu}$. For the coefficients $P_{2\nu}$ we have

$$P_2 = \frac{28}{45}, \quad P_4 = -\frac{6706}{151875}, \quad P_6 = \frac{1163716}{205031250}.$$

We now form the following table

ν	$15\nu + 207$	$\frac{1}{15\nu + 207}$	δ_{15}	δ^2_{15}	δ^3_{15}
−2	177	0.005649718			
−1	192	5208333	−441385	63970	
0	207	4830918	−377415	51002	−12968
1	222	4504505	−326413		
2	237	4219409			
3	252	3968254			
4	267	3745318			
5	282	3546099			
6	297	3367003			
7	312	3205128			
8	327	3058104			
9	342	2923977	−134127	11270	
10	357	2801120	−122857	9909	−1361
11	372	2688172	−112948		

$$\sum_{\nu=0}^{9} = 0.037368715$$

and obtain by (11), apart from the remainder-term,

$$\sum_{200}^{349} \frac{1}{\nu} = 15 \times 0.037368715 + P_2(-0.000122857 + 0.000377415)$$
$$+ P_4(-0.000001361 + 0.000012968)$$
$$= 0.560530725 + 0.000158392 - 0.000000513$$
$$= 0.56068860.$$

For the remainder-term we find[1]

$$R = 10P_6 15^6 \frac{6\,!}{(200+\xi)^7} < 0.0^7 86,$$

so that the error does not exceed 9 units of the 8th decimal. The correct value to 8 decimals is, in fact, 0.56068862.

If we calculate the same sum by the second formula of Woolhouse's type, (21), the principal term remains unaltered, while for the corrective terms we find

$$\frac{1}{12}\frac{15^2-1}{2\,!}\left(\frac{1}{199.5^2} - \frac{1}{349.5^2}\right) - \frac{7}{240}\frac{15^4-1}{4\,!}\left(\frac{6}{199.5^4} - \frac{6}{349.5^4}\right)$$
$$= 0.00015789,$$

so that the result becomes 0.56068861. This result can only be affected by forcing-errors, as we find, by (26), for the remainder-term

$$R = -\frac{B_6(\frac{1}{2})}{6\,!} 150(15^6 - 1) \frac{6\,!}{(200+\xi)^7} < 0.0^8 31.$$

166. Whether, under given circumstances, Lubbock's or Woolhouse's type is to be preferred, does not entirely depend on the accuracy obtainable by a given number of terms—in which respect Woolhouse's type is preferable—but also on the difficulty involved in calculating the differential coefficients. If the numerical values of the integral of the function are easily accessible, neither of these formulas should be applied, but a formula of Gauss's or Euler's type.

[1] Instead of n zeros in succession we shall sometimes, for abbreviation, write 0^n.

§16. Mechanical Quadrature

167. By a formula of mechanical quadrature, we[1] mean a formula expressing the approximate value of an integral as a linearfunction of a certain number of values of the function to be integrated. Formulas of this nature may be obtained in many different ways; here, we confine ourselves to considering certain formulas obtained by integrating Lagrange's interpolation-formula between definite limits. If these are fixed, for instance 0 and 1, and the arguments $a_0, a_1, \ldots a_n$ also, for instance $0, \frac{1}{n}, \frac{2}{n}, \ldots \frac{n-1}{n}, 1$, we may once and for all calculate the coefficients of the values of $f(x)$ employed in the formula (Cotes' method). As we may always, by a linear transformation, ensure that the limits of integration become given numbers, we may at any step of our calculations use the limits that lead to the simplest operations.

168. We content ourselves with deducing certain quadrature formulas for which it is possible to give the remainder-term a practical form.

Using first for the interpolation the points $0, \pm 1, \pm 2, \ldots \pm r$, we put[2]

$$P(x) = x(x^2 - 1)(x^2 - 4) \ldots (x^2 - r^2) \tag{1}$$

and

$$P_\nu(x) = \frac{P(x)}{x - \nu}, \tag{2}$$

ν being one of the values $0, \pm 1, \pm 2, \ldots \pm r$, and find, by Lagrange's interpolation formula

$$f(x) = \sum_{-r}^{r} \frac{P_\nu(x)}{P_\nu(\nu)} f(\nu) + P(x) f(x, 0, \pm 1, \ldots \pm r). \tag{3}$$

Here, $f(x, 0, \pm 1, \ldots \pm r)$ means the divided difference of order $2r + 1$, formed with the arguments $x, 0, \pm 1, \ldots \pm r$.

[1] Many authors use the expression as comprising all formulas which can be used for numerical integration, comprising also the summation-formulas of Laplace's and Euler's types.

[2] In the case of $r = 0$, we put $P(x) = x$.

We now put, m denoting a positive integer,

$$U_\nu = \int_{-m}^{m} \frac{P_\nu(x)}{P_\nu(\nu)} dx. \tag{4}$$

As $P(-x) = -P(x)$, we have

$$P_\nu(-x) = \frac{P(-x)}{-x-\nu} = \frac{P(x)}{x+\nu}$$

or $P_\nu(-x) = P_{-\nu}(x)$. By means of this relation, we find

$$U_{-\nu} = \int_{-m}^{m} \frac{P_{-\nu}(x)}{P_{-\nu}(-\nu)} dx = \int_{-m}^{m} \frac{P_\nu(-x)}{P_\nu(\nu)} dx,$$

and therefore $U_{-\nu} = U_\nu$, as is seen by putting $x = -t$ in the last integral.

Integrating (3) between the limits[1] $\pm m$, we consequently find

$$\int_{-m}^{m} f(x)dx = U_0 f(0) + \sum_1^r U_\nu[f(\nu) + f(-\nu)] + R, \tag{5}$$

$$R = \int_{-m}^{m} P(x)f(x, 0, \pm 1, \ldots \pm r)dx. \tag{6}$$

169. This remainder-term may be simplified by partial integration, but it is first necessary to examine the function

$$Q(x) = \int_{-m}^{x} P(t)dt. \tag{7}$$

We begin by observing that $Q(-m) = 0$ and $Q(m) = 0$; the latter result following from $P(-t) = -P(t)$.

Next, we may prove that $Q(x)$ *does not vanish in the interval* $\pm m$. If we put

$$I_\nu = \int_\nu^{\nu+1} P(t)dt, \tag{8}$$

[1] The case where the limits of integration are $\pm(m + \frac{1}{2})$ has been treated recently by A. Walther: Zur numerischen Integration. Skandinavisk Aktuarietidskrift 1925, p. 148.

the assertion will evidently be proved if the graph of $P(t)$ is as indicated in figure 5,

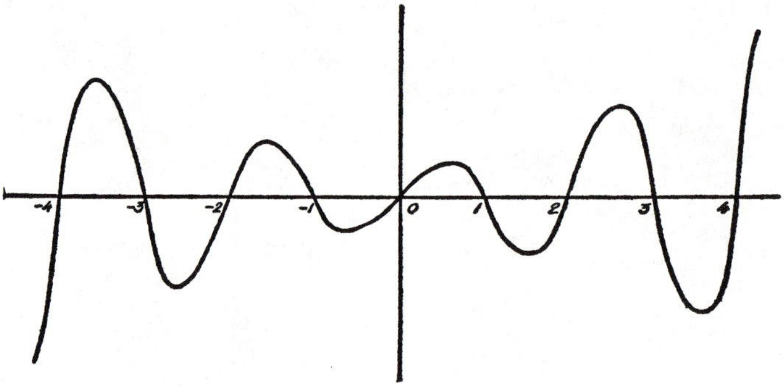

FIG. 5

that is, if $I_{-\nu} = -I_{\nu-1}$, while the sequence $I_0, I_1, I_2, \ldots$, and consequently also the sequence $I_{-1}, I_{-2}, I_{-3}, \ldots$, is constantly increasing in absolute value.

Now we have

$$I_{-\nu} = \int_{-\nu}^{-\nu+1} P(t)dt = \int_{\nu-1}^{\nu} P(-t)dt = -\int_{\nu-1}^{\nu} P(t)dt = -I_{\nu-1}.$$

Next, if in (8) we introduce $P(t) = (t + r)^{(2r+1)}$, we find

$$I_{\nu-1} = \int_{\nu-1}^{\nu} (t + r)^{(2r+1)}dt = \int_{\nu}^{\nu+1} (t + r - 1)^{(2r+1)}dt$$

$$= \int_{\nu}^{\nu+1} \frac{t - r - 1}{t + r} (t + r)^{(2r+1)}dt$$

or

$$I_{\nu-1} = \int_{\nu}^{\nu+1} \frac{t - r - 1}{t + r} P(t)dt. \tag{9}$$

As $P(t)$ does not vanish in the interval from ν to $\nu + 1$, we may apply the Theorem of Mean Value to this integral and get

$$I_{\nu-1} = \frac{\xi - r - 1}{\xi + r} I_\nu \qquad (\nu < \xi < \nu + 1). \tag{10}$$

But for $\nu > r$ we have $0 < \xi - r - 1 < \xi + r$, and for $0 < \nu \leq r$ we have $0 < r + 1 - \xi < \xi + r$, so that $|I_{\nu-1}| < |I_\nu|$ for $\nu > 0$, as we desired to prove.

170. Making use of the properties of $Q(x)$ just proved, we may transform (6) by partial integration, and by means of the Theorem of Mean Value. We find in this way

$$R = -\int_{-m}^{m} Q(x) f(x, x, 0, \pm 1, \ldots \pm r)\, dx$$

$$= -f(\xi, \xi, 0, \pm 1, \ldots \pm r) \int_{-m}^{m} Q(x)\, dx$$

$$= -\frac{f^{(2r+2)}(\xi)}{(2r+2)!} \int_{-m}^{m} Q(x) dx;$$

but we have by partial integration

$$\int_{-m}^{m} Q(x)dx = \Big[xQ(x)\Big]_{-m}^{m} - \int_{-m}^{m} xQ'(x)dx$$

$$= -\int_{-m}^{m} xP(x)dx,$$

so that finally, as $xP(x) = x^{[2r+2]}$,

$$R = \frac{2f^{(2r+2)}(\xi)}{(2r+2)!} \int_{0}^{m} x^{[2r+2]}dx. \tag{11}$$

171. If unity is chosen as the interval of integration, we must, in (5), put $f(x) = F\left(\frac{x}{2m}\right)$. If at the same time we write, for abbreviation,

$$V_\nu = \frac{1}{2m} U_\nu,\ F_{\pm\nu} = F\left(\frac{\nu}{2m}\right) + F\left(-\frac{\nu}{2m}\right), \tag{12}$$

the formula becomes

$$\left.\begin{aligned} \int_{-\frac{1}{2}}^{\frac{1}{2}} F(x)dx &= V_0F_0 + \sum_{1}^{r} V_\nu F_{\pm\nu} + R, \\ R &= O_{2m}^{[2r+2]} F^{(2r+2)}(\xi), \end{aligned}\right\} \tag{13}$$

where we have put

$$
\left.\begin{aligned}
O_{2m}^{[2k]} &= \frac{2}{(2k)!\,(2m)^{2k+1}} \int_0^m x^{[2k]}dx \\
&= \frac{1}{(2k)!\,(2m)^{2k}} \sum_{\nu=1}^{k} m^{2\nu} \frac{D^{2\nu}0^{[2k]}}{(2\nu+1)!}.
\end{aligned}\right\} \quad (14)
$$

A partial check on the calculation of the coefficients V_ν is obtained by putting $F(x) = 1$ whence

$$V_0 + 2\sum_{1}^{r} V_\nu = 1.$$

The coefficient of the remainder-term is controlled by putting $F(x) = x^{2r+2}$.

172. The two most important cases of (13) are obtained for $r = m$ and $r = m - 1$. They are called the *closed* and the *open type* respectively, because the values of $F(x)$ at the end points of the interval are used in the former, but not in the latter case. The resulting formulas with 3, 5, 7, 9 and 11 terms may immediately be written down by means of the tables given below.

Quadrature-Formulas of the Closed Type

NUMBER OF TERMS	F_0	$F_{\pm 1}$	$F_{\pm 2}$	$F_{\pm 3}$	$F_{\pm 4}$	$F_{\pm 5}$	COMMON DIVISOR	REMAINDER-TERM
3	4	1					6	$-0.0^3\,35F^{(4)}\,(\xi)$
5	12	32	7				90	$-0.0^6\,52F^{(6)}\,(\xi)$
7	272	27	216	41			840	$-0.0^9\,64F^{(8)}\,(\xi)$
9	−4540	10496	−928	5888	989		28350	$-0.0^{12}59F^{(10)}\,(\xi)$
11	427368	−260550	272400	−48525	106300	16067	598752	$-0.0^{15}42F^{(12)}\,(\xi)$

Quadrature-Formulas of the Open Type

NUMBER OF TERMS	F_0	$F_{\pm 1}$	$F_{\pm 2}$	$F_{\pm 3}$	$F_{\pm 4}$	$F_{\pm 5}$	COMMON DIVISOR	REMAINDER-TERM
3	−1	2					3	$0.0^3\,31F^{(4)}\,(\xi)$
5	26	−14	11				20	$0.0^5\,11F^{(6)}\,(\xi)$
7	−2459	2196	−954	460			945	$0.0^8\,21F^{(8)}\,(\xi)$
9	67822	−55070	33340	−11690	4045		9072	$0.0^{11}27F^{(10)}\,(\xi)$
11	−494042	427956	−266298	123058	−35771	9626	23100	$0.0^{14}25F^{(12)}\,(\xi)$

To these may, as regards the open type, be added the one-term formula

$$\int_{-\frac{1}{2}}^{\frac{1}{2}} F(x)dx = F(0) + \frac{F''(\xi)}{24}. \tag{15}$$

The tables need hardly any explanation. The columns headed $F_{\pm\nu}$ contain the numerator of V_ν while the denominator, being the same for each line, is stated in the column headed "Common divisor."

As an example, the five-term formula of the closed type is, written in full,

$$\int_{-\frac{1}{2}}^{\frac{1}{2}} F(x)dx = \frac{1}{90}(12F_0 + 32F_{\pm1} + 7F_{\pm2}) - 0.0^652F^{(6)}(\xi) \tag{16}$$

and the corresponding formula of the open type

$$\int_{-\frac{1}{2}}^{\frac{1}{2}} F(x)dx = \frac{1}{20}(26F_0 - 14F_{\pm1} + 11F_{\pm2}) + 0.0^511F^{(6)}(\xi). \tag{17}$$

It should be carefully noted that only F_0, being the middle-point value in both cases, has the same meaning in the two formulas. $F_{\pm2}$ means the sum of the end-point values in (16), the type being closed, but not in (17), the type being open.

The constants of the remainder-terms have in the tables been stated with approximate and slightly too large values, ready for the practical applications. For instance

$$0.0^335 = 0.00035 > \frac{1}{2880}.$$

The exact values of these constants are, in the case of the closed type,

$$\frac{1}{2880},\ \frac{1}{1935360},\ \frac{1}{1567641600},\ \frac{37}{62783697715200},$$

$$\frac{26927}{653837184000000000},$$

the signs being all negative; and in the case of the open type,

$$\frac{7}{23040},\ \frac{41}{39191040},\ \frac{989}{475634073600},\ \frac{16067}{5987520000000000},$$

$$\frac{1364651}{5622760425683681280000},$$

the signs being all positive.

The three-term formula of the closed type, or

$$\int_{-\frac{1}{2}}^{\frac{1}{2}} F(x)dx = \frac{1}{6}(4F_0 + F_{\pm 1}) - 0.0^335F^{(4)}(\xi) \qquad (18)$$

has a special name and is called Simpson's formula. It must be compared with the three-term formula of the open type, or

$$\int_{-\frac{1}{2}}^{\frac{1}{2}} F(x)dx = \frac{1}{3}(-F_0 + 2F_{\pm 1}) + 0.0^331F^{(4)}(\xi). \qquad (19)$$

It is rather striking that the latter formula is slightly more accurate than Simpson's formula, although a comparison between the formulas of the closed and the open type is generally to the advantage of the former.

The formulas of the open type are important on account of their application to the numerical integration of differential equations. But this subject will be dealt with separately in §17.

173. The formulas of mechanical quadrature are particularly adapted to the arithmometer, but are also easy to handle with a product-table carried to two or three figures, especially if the number of terms in the formula is small. But as the number of terms increases, the coefficients become more inconvenient. It is, therefore, sometimes preferable to apply one of the simpler formulas several times to minor parts of the given interval, rather than to apply a formula of higher order once to the whole interval. This method of proceeding, however, can as a rule only be recommended when the required values of the function are at hand, or at least easy to calculate; as a formula of this nature, generally speaking, does not yield as accurate a result as the formula of higher order with the same number of terms.

As an example, if we apply Simpson's formula to the two intervals $(-\frac{1}{2}, 0)$ and $(0, \frac{1}{2})$, and add the results, we get

$$\int_{-\frac{1}{2}}^{\frac{1}{2}} F(x)dx = \frac{1}{12}(2F_0 + 4F_{\pm 1} + F_{\pm 2}) - \frac{F^{(4)}(\xi)}{46080}. \tag{20}$$

It is at once seen that this formula, using the same values of $F(x)$ as in the five-term formula of the closed type, is as a rule far inferior to the latter. An exception must be made, if $F^{(6)}(\xi)$ can attain such far greater values than $F^{(4)}(\xi)$ that (20) is, for that reason, preferable.

174. As an example, let us calculate

$$\int_{100000}^{200000} \frac{dx}{\operatorname{Log} x}$$

by means of (16). The data are given in the table below. In this we have stated the required values of $\log x$ to six places and taken

x	$\log x$	$\frac{1}{\log x}$
100000	5.000000	0.2000000
125000	5.096910	0.1961973
150000	5.176091	0.1931960
175000	5.243038	0.1907291
200000	5.301030	0.1886426

out the reciprocals to seven places. It is preferable to carry out the calculation with Brigg's logarithms, and convert to natural logarithms at the end of the calculation.

We now have, apart from the remainder-term,

$$\int_{100000}^{200000} \frac{dx}{\log x} = \frac{100000}{90}(12\times 0.1931960 + 32\times 0.3869264 + 7\times 0.3886426)$$

$$= 19356.11,$$

whence, on multiplying by $\log e = 0.43429448$,

$$\int_{100000}^{200000} \frac{dx}{\operatorname{Log} x} = 8406.25.$$

As regards the remainder-term, we find (see, for instance, §21, formula (35))

$$D^6 \frac{1}{\operatorname{Log} x} = \frac{1}{x^6}\left(\frac{120}{\operatorname{Log}^2 x} + \frac{548}{\operatorname{Log}^3 x} + \frac{1350}{\operatorname{Log}^4 x} + \frac{2040}{\operatorname{Log}^5 x} + \frac{1800}{\operatorname{Log}^6 x} + \frac{720}{\operatorname{Log}^7 x}\right).$$

Putting $x = 100000$, we find, as $\operatorname{Log} 100000 = 11.513$,

$$D^6 \frac{1}{\operatorname{Log} \xi} < \frac{1.352}{100000^6},$$

so that, taking the unit interval into account, we obtain

$$|R| < 0.0^652 \times 1.352 \times 100000 < 0.07\dot{1}.$$

The remainder-term being negative, the value found for the integral is slightly too large.

175. In deriving the remainder-terms for the quadrature-formulas above, it was of vital importance that the function $Q(x)$ should not vanish within the interval of integration. The same method is, therefore, not applicable when the number of values of the function is *even*. In this case, we proceed as follows.

If we put

$$P(x) = (x^2 - \tfrac{1}{4})(x^2 - \tfrac{9}{4}) \ldots \left[x^2 - \frac{(2r-1)^2}{4}\right], \tag{21}$$

$$P_\nu(x) = \frac{P(x)}{x - \dfrac{2\nu - 1}{2}}, \tag{22}$$

ν being one of the numbers $0, \pm 1, \pm 2, \ldots \pm(r-1), r$, we find by Lagrange's formula

$$f(x) = \sum_{-r+1}^{r} \frac{P_\nu(x)}{P_\nu\left(\dfrac{2\nu-1}{2}\right)} f\left(\frac{2\nu-1}{2}\right) + P(x) f\left(x, \pm\tfrac{1}{2}, \ldots \pm\frac{2r-1}{2}\right). \tag{23}$$

We now put, m being positive and integral,

$$U_\nu = \int_{-m-\frac{1}{2}}^{m+\frac{1}{2}} \frac{P_\nu(x)}{P_\nu\left(\dfrac{2\nu - 1}{2}\right)} dx. \tag{24}$$

If we note that

$$P(-x) = P(x),\ P_\nu(-x) = -P_{-\nu+1}(x),$$

we may prove, as in No. 168, that $U_\nu = U_{-\nu+1}$, and thus obtain

$$\int_{-m-\frac{1}{2}}^{m+\frac{1}{2}} f(x)dx = \sum_1^r U_\nu\left[f\left(\frac{2\nu-1}{2}\right) + f\left(-\frac{2\nu-1}{2}\right)\right] + R, \quad (25)$$

$$R = \int_{-m-\frac{1}{2}}^{m+\frac{1}{2}} P(x) f\left(x, \pm\tfrac{1}{2}, \ldots \pm \frac{2r-1}{2}\right)dx. \quad (26)$$

176. In order to simplify this remainder-term, we integrate first from $-m-\frac{1}{2}$ to $m-\frac{1}{2}$, and then from $m-\frac{1}{2}$ to $m+\frac{1}{2}$. To the latter part of the integral we may apply the Theorem of Mean Value, as $P(x)$ does not change its sign here. We thus obtain

$$R = \int_{-m-\frac{1}{2}}^{m-\frac{1}{2}} P(x) f\left(x, \pm\tfrac{1}{2}, \ldots \pm \frac{2r-1}{2}\right)dx + \frac{f^{(2r)}(\xi_1)}{(2r)!}\int_{m-\frac{1}{2}}^{m+\frac{1}{2}} P(x)dx. \quad (27)$$

Now we have, by §3 (1),

$$\left(x - \frac{2r-1}{2}\right) f\left(x, \pm\tfrac{1}{2}, \ldots \pm \frac{2r-1}{2}\right) =$$

$$f\left(x, \pm\tfrac{1}{2}, \ldots \pm \frac{2r-3}{2}, -\frac{2r-1}{2}\right) - f\left(\pm\tfrac{1}{2}, \ldots \pm \frac{2r-1}{2}\right),$$

and as evidently $P(x) = (x + r - \frac{1}{2})^{(2r)}$, we find for the first integral in (27)

$$\int_{-m-\frac{1}{2}}^{m-\frac{1}{2}} (x + r - \tfrac{1}{2})^{(2r-1)} f\left(x, \pm\tfrac{1}{2}, \ldots \pm \frac{2r-3}{2}, -\frac{2r-1}{2}\right) -$$

$$f\left(\pm\tfrac{1}{2}, \ldots \pm \frac{2r-1}{2}\right)\int_{-m-\frac{1}{2}}^{m-\frac{1}{2}} (x + r - \tfrac{1}{2})^{(2r-1)} dx.$$

If, in both of these integrals, we replace x by $x - \frac{1}{2}$, we see that the latter vanishes, while the former may be written

$$\int_{-m}^{m} x(x^2-1)\ldots[x^2-(r-1)^2] f\left(x-\tfrac{1}{2}, \pm\tfrac{1}{2}, \ldots \pm \frac{2r-3}{2}, -\frac{2r-1}{2}\right)dx.$$

But this integral may be treated by the same method which we applied in order to derive (11) from (6). We may thus immediately see, without having actually to perform the calculations again, that the above integral may be written

$$\frac{2f^{(2r)}(\xi_2)}{(2r)!}\int_0^m x^{[2r]}dx,$$

so that the remainder-term assumes the form

$$R = \frac{2f^{(2r)}(\xi_2)}{(2r)!}\int_0^m x^{[2r]}dx + \frac{f^{(2r)}(\xi_1)}{(2r)!}\int_{m-\frac{1}{2}}^{m+\frac{1}{2}} x^{[2r+1]-1}dx. \tag{28}$$

The two terms of which this remainder-term is composed, may be contracted into one. We first note that

$$\int_0^t x^{[2r]}dx = \frac{1}{2r+1}\int_{t-\frac{1}{2}}^{t+\frac{1}{2}} x^{[2r+1]}\,dx; \tag{29}$$

for both sides vanish for $t = 0$, and their differential coefficients are $t^{[2r]}$ and $\frac{1}{2r+1}\,\delta t^{[2r+1]} = t^{[2r]}$ respectively. We may, therefore, write (28) in the form

$$R = \frac{2f^{(2r)}(\xi_2)}{(2r+1)!}\int_{m-\frac{1}{2}}^{m+\frac{1}{2}} x^{[2r+1]}dx + \frac{f^{(2r)}(\xi_1)}{(2r)!}\int_{m-\frac{1}{2}}^{m+\frac{1}{2}} x^{[2r+1]-1}\,dx;$$

but here the coefficients of $f^{(2r)}(\xi_1)$ and $f^{(2r)}(\xi_2)$ have *the same sign*, so that we may put

$$R = \frac{2f^{(2r)}(\xi)}{(2r+1)!}\int_{m-\frac{1}{2}}^{m+\frac{1}{2}} [x^{[2r+1]} + (r+\tfrac{1}{2})x^{[2r+1]-1}]dx$$

$$= \frac{2f^{(2r)}(\xi)}{(2r+1)!}\int_{m-\frac{1}{2}}^{m+\frac{1}{2}} (x+r+\tfrac{1}{2})^{(2r+1)}dx$$

or

$$R = \frac{2f^{(2r)}(\xi)}{(2r+1)!}\int_{m}^{m+1} x^{[2r+2]-1}dx. \tag{30}$$

From this formula, we may find the sign of the coefficient of $f^{(2r)}(\xi)$, as also, from (29), we may find the sign of the coefficient in

(11). We may also by means of these formulas find limits between which these coefficients are comprised. We leave, however, these details to the reader.

The remainder-term (30) may be put into a form which is more convenient for the calculation. We have

$$\int_t^{t+1} x^{[2r+2]-1}dx = (2r+1)\int_0^{t+\frac{1}{2}} x^{[2r+1]-1}dx; \tag{31}$$

for both sides vanish for $t = -\frac{1}{2}$, and their differential coefficients are

$$(t+1)^{[2r+2]-1} - t^{[2r+2]-1} = \delta(t+\tfrac{1}{2})^{[2r+2]-1}$$

and

$$(2r+1)\,(t+\tfrac{1}{2})^{[2r+1]-1}$$

which, according to §2(10), are identical. We may therefore write (30) in the form

$$R = \frac{2f^{(2r)}(\xi)}{(2r)!}\int_0^{m+\frac{1}{2}} x^{[2r+1]-1}dx. \tag{32}$$

177. It should be observed that this form of the remainder-term is the same as that which would result if the Theorem of Mean Value were applicable to (26). This must not, however, be taken as a generalization of the Theorem of Mean Value; for we have *not* proved that

$$R = f\left(\eta, \pm\tfrac{1}{2}, \ldots \pm\frac{2r-1}{2}\right)\int_{-m-\frac{1}{2}}^{m+\frac{1}{2}} P(x)dx.$$

It is, therefore, possible that no η exists, satisfying the latter relation, although it is certain that a ξ exists, satisfying (32).

178. If we choose the interval of integration as unity, we must, in (25), put $f(x) = F\left(\frac{x}{2m+1}\right)$. If, at the same time, we introduce the notations

$$V_\nu = \frac{1}{2m+1}U_\nu, \quad F_{\pm\nu} = F\left(\frac{\nu-\frac{1}{2}}{2m+1}\right) + F\left(-\frac{\nu-\frac{1}{2}}{2m+1}\right), \tag{33}$$

the formula becomes

$$\int_{-\frac{1}{2}}^{\frac{1}{2}} F(x)dx = \sum_{1}^{r} V_\nu F_{\pm\nu} + R, \tag{34}$$

$$R = O_{2m+1}^{[2r+1]-1} F^{(2r)}(\xi), \tag{35}$$

putting

$$\left.\begin{aligned} O_{2m+1}^{[2r+1]-1} &= \frac{2}{(2r)!\,(2m+1)^{2r+1}} \int_0^{m+\frac{1}{2}} x^{[2r+1]-1} dx \\ &= \frac{1}{(2r)!\,(2m+1)^{2r}} \sum_{\nu=0}^{r} \frac{(2m+1)^{2\nu}}{2^{2\nu}(2\nu+1)} \frac{D^{2\nu+1}0^{[2r+1]}}{(2\nu+1)!} \end{aligned}\right\} \tag{36}$$

179. Here, too, we obtain a closed and an open type, the *closed* one for $r = m + 1$, the *open* one for $r = m$. We find for the first few values of $-O_{2r-1}^{[2r+1]-1}$, commencing with $r = 1$,

$$\frac{1}{12}, \frac{1}{6480}, \frac{11}{37800000}, \frac{167}{426924691200}, \frac{173}{458209960750080},$$

and for $O_{2r+1}^{[2r+1]-1}$

$$\frac{1}{36}, \frac{19}{90000}, \frac{751}{1016487360}, \frac{2857}{1928493100800}, \frac{434293}{2258921433341061120}.$$

The resulting formulas with 2, 4, 6, 8 and 10 terms may be written down by means of the following tables, which have been arranged on the same principle as the tables for the case of an odd number of terms. A comparison with the latter shows that nothing much

Quadrature-Formulas of the Closed Type

NUMBER OF TERMS	$F_{\pm 1}$	$F_{\pm 2}$	$F_{\pm 3}$	$F_{\pm 4}$	$F_{\pm 5}$	COMMON DIVISOR	REMAINDER-TERM
2	1					2	$-\frac{1}{12}F^{(2)}(\xi)$
4	3	1				8	$-0.0^3\,16F^{(4)}(\xi)$
6	50	75	19			288	$-0.0^6\,30F^{(6)}(\xi)$
8	2989	1323	3577	751		17280	$-0.0^9\,40F^{(8)}(\xi)$
10	5778	19344	1080	15741	2857	89600	$-0.0^{12}38F^{(10)}(\xi)$

Quadrature-Formulas of the Open Type

NUMBER OF TERMS	$F_{\pm 1}$	$F_{\pm 2}$	$F_{\pm 3}$	$F_{\pm 4}$	$F_{\pm 5}$	COMMON DIVISOR	REMAINDER-TERM
2	1					2	$\frac{1}{36}F^{(2)}\ (\xi)$
4	1	11				24	$0.0^3\,22F^{(4)}\ (\xi)$
6	562	−453	611			1440	$0.0^6\,74F^{(6)}\ (\xi)$
8	−1711	4967	−2803	1787		4480	$0.0^8\,15F^{(8)}\ (\xi)$
10	8891258	−17085616	15673880	−6603199	2752477	7257600	$0.0^{11}20F^{(10)}\ (\xi)$

is gained by using a term more if, by so doing, the number of terms becomes even. The formulas with an even number of terms should, therefore, as a rule only be used, when the table at hand does not contain the values required in the formulas with an odd number of terms.

180. Any number of variants of the quadrature-formulas dealt with above may be obtained by adding to the formula $K\delta^m f(0)$, K being a constant, and $\delta^m f(0)$ being expressed as a linear function of the values of $f(x)$ employed in the difference. In that case, the remainder-term must be corrected by deducting $K\delta^m f(0) = Kf^{(m)}(\eta)$. While most of these variants are without interest, one or two of them have obtained practical importance, either because their coefficients are simpler, or because their accuracy is greater, than in the case of Cotes' formula with the same number of terms. The latter case may occur, if the constant K is chosen in such a way that a term in the formula is caused to vanish.

As an example, let us consider the seven-term formula of the closed type which, if the interval of integration is taken as 6, assumes the form

$$\int_{-3}^{3} f(x)dx = \frac{1}{140}(272f_0 + 27f_{\pm 1} + 216f_{\pm 2} + 41f_{\pm 3}) - \frac{9}{1400}f^{(8)}(\xi), \quad (37)$$

and let us from this deduct

$$\frac{9}{700}\,\delta^6 f(0) = \frac{9}{700}(-20f_0 + 15f_{\pm 1} - 6f_{\pm 2} + f_{\pm 3}),$$

adding, at the same time, $\frac{9}{700}f^{(6)}(\eta)$ to the remainder-term. The result is Hardy's formula

$$\int_{-3}^{3} f(x)dx = 2.2f_0 + 1.62f_{\pm 2} + 0.28f_{\pm 3} + R, \tag{38}$$

$$R = \frac{9}{700}[f^{(6)}(\eta) - \tfrac{1}{2}f^{(8)}(\xi)]. \tag{39}$$

It is natural to ask whether the remainder-term may not be simplified by putting

$$f^{(6)}(\eta) - \tfrac{1}{2}f^{(8)}(\xi) = f^{(6)}(\zeta).$$

This is, however, not possible. It is, in fact, easy to construct a function $f(x)$ of such a nature that $f^{(8)}(x)$, while continuous in the closed interval ± 3, assumes very large values therein, while $f^{(6)}(x)$ only assumes moderate values.

It is seen that the term $f_{\pm 1}$ has disappeared in Hardy's formula, and it must therefore be compared with the five-term formula (16). If, now, we transform the interval of integration in (38) so that it becomes $\pm\frac{1}{2}$, Hardy's formula may be written

$$\int_{-\frac{1}{2}}^{\frac{1}{2}} F(x)dx = \frac{1}{3}(1.1F_0 + 0.81F_{\pm 2} + 0.14F_{\pm 3}) + R, \tag{40}$$

$$R = \frac{F^{(6)}(\eta) - \frac{1}{72}F^{(8)}(\xi)}{21772800}. \tag{41}$$

In the frequently occurring cases where $\frac{1}{72}F^{(8)}(\xi)$ is negligible in comparison with $F^{(6)}(\eta)$, we therefore, on the average, gain about one decimal in accuracy by using Hardy's instead of Cotes' formula. On the other hand, the examination of the remainder-term is slightly more elaborate in the case of Hardy's formula.

181. Another favourite formula is obtained if, in (37), we add $\frac{1}{140}\delta^6 f(0)$. The result is Weddle's formula

$$\int_{-3}^{3} f(x)dx = \frac{3}{10}(6f_0 + f_{\pm 1} + 5f_{\pm 2} + f_{\pm 3}) + R, \tag{42}$$

$$R = -\frac{1}{140}[f^{(6)}(\eta) + 0.9f^{(8)}(\xi)]. \tag{43}$$

In this case no term vanishes, and the formula must therefore be compared with the seven-term formula (37). It is seen that Weddle's formula owes its popularity to the simplicity of the coefficients, and that (37) is more accurate when $f^{(8)}(\xi)$ is numerically smaller than $f^{(6)}(\eta)$.

182. Another type of formulas of mechanical quadrature, where the remainder-term assumes a practical form, is obtained by §4 (35). In this case it is assumed that both $f(x)$ and $f'(x)$ are known at the points a_ν.

We put

$$P(x) = (x - a_0)(x - a_1) \ldots (x - a_n), \quad P_\nu(x) = \frac{P(x)}{x - a_\nu}, \tag{44}$$

so that

$$P_\nu(a_\nu) = P'(a_\nu), \quad P_\nu'(a_\nu) = \tfrac{1}{2}P''(a_\nu); \tag{45}$$

further

$$\left.\begin{aligned} W_\nu' &= \int_a^b \frac{P_\nu(x)P(x)}{P_\nu^2(a_\nu)}dx, \\ W_\nu &= \int_a^b \frac{P_\nu^2(x)}{P_\nu^2(a_\nu)}dx - 2W_\nu' \frac{P_\nu'(a_\nu)}{P_\nu(a_\nu)}, \end{aligned}\right\} \tag{46}$$

and obtain from §4 (35) by integration from a to b

$$\int_a^b f(x)dx = \sum_0^n [W_\nu f(a_\nu) + W_\nu' f'(a_\nu)] + R, \tag{47}$$

and for the remainder-term, as $P^2(x)$ does not change its sign,

$$\left.\begin{aligned} R &= \int_a^b P^2(x) f(x, a_0, a_0, \ldots a_n, a_n)dx \\ &= \frac{f^{(2n+2)}(\xi)}{(2n+2)!}\int_a^b P^2(x)dx. \end{aligned}\right\} \tag{48}$$

The most accurate formula of this type is obtained, if the numbers a_ν are determined in such a way that the coefficient of $f^{(2n+2)}(\xi)$

in the remainder-term becomes a minimum. We must, then, for $\nu = 0, 1, 2, \ldots n$ have

$$\frac{\partial}{\partial a_\nu}\int_a^b P^2(x)dx = 0$$

or

$$\int_a^b P_\nu(x)P(x)dx = 0 \quad (\nu = 0, 1, \ldots n).$$

It follows that $W_\nu' = 0$, so that (47) is considerably simplified, as the terms implying $f'(a_\nu)$ vanish. The values of a_ν and W_ν resulting for $a = 0$, $b = 1$, have been calculated by Gauss and are, for instance, stated in Markoff's "Differenzenrechnung," p. 70. We shall, however, not go into details about this equally interesting and accurate method, because its practical applicability is hampered by the fact that it leads to inconvenient (generally irrational) values for a_ν and W_ν.

The Gaussian formula forms the object of a large amount of literature. We refer the reader, both as regards this subject, and concerning numerical integration in general, to the article by Runge & Willers in "Encyklopädie der Mathematischen Wissenschaften II, C. 2., p. 47.[1]

§17. Numerical Integration of Differential Equations

183. In order to show how it is possible, in principle, by means of numerical methods to approximate to the integral of a given differential equation, we commence by considering a differential equation of the first order

$$\frac{dx}{dt} = f(x, t) \tag{1}$$

and try to determine the integral which for $t = t_0$ assumes the value $x = x_0$. Previously it must, of course, be ascertained that this integral exists, which is here taken for granted.

[1] See also a recent paper by E. Lefrancq: De l'Interpolation (Bulletin du comité permanent des Congrès Internationaux d'Actuaires, No. 21).

It follows from the given conditions that

$$x = x_0 + \int_{t_0}^{t} f(x, t)\, dt; \tag{2}$$

in integrating on the right it must be taken into account that x is a function of t.

We assume $t > t_0$ and divide the interval from t_0 to t into n equal parts h, so that $t = t_0 + nh$. If, for a moment, we assume that *the value of x is already known for the following values of t*

$$t_0 + h,\ t_0 + 2h,\ \ldots,\ t_0 + (n-1)h, \tag{3}$$

the problem is simply to calculate the integral from t_0 to $t_0 + nh$ of a function whose values are known for the values (3) of the variable, and to indicate limits to the error committed. But this problem can be solved by a simple application of the quadrature-formulas of the open type, given in §16, having previously transformed the interval to suit the problem under consideration.

By these simple considerations the problem is, in principle, solved if we know the values of x corresponding to the initial values (3) of the argument; but in order to start the calculation at all, we must first procure these. In what way this is done, is indifferent; we may, for instance, in many cases start the calculation at a point where it is possible to develop $(x - x_0)$ in powers of $(t - t_0)$ in a rapidly convergent series.

184. If we have to deal with a differential equation of higher order than the first, we may commence by putting it into the form of a *differential system*

$$\frac{dx_i}{dt} = f_i(x_1, x_2, \ldots x_k, t) \qquad (i = 1, 2, \ldots k), \tag{4}$$

this being—as is well known—always possible by introducing auxiliary variables. We content ourselves with considering a system of two such equations, as the extension to any number of equations does not present any difficulties whatever.

Let, then, the system of two differential equations be

$$\left.\begin{aligned} x' &= f(x, y, t) \\ y' &= \varphi(x, y, t) \end{aligned}\right\} \tag{5}$$

or, if we consider the integral that for $t = t_0$ assumes the values $x = x_0, y = y_0$, a system of two *integral equations*

$$\left.\begin{aligned} x - x_0 &= \int_{t_0}^{t} f(x, y, t)dt, \\ y - y_0 &= \int_{t_0}^{t} \varphi(x, y, t)dt. \end{aligned}\right\} \quad (6)$$

It is, as before, assumed that the existence of the integral has been previously ascertained.

We now assume that $x(t)$ and $y(t)$ have already been tabulated up to and including t; at the start, this may, for instance, have been done by expanding $(x - x_0)$ and $(y - y_0)$ in powers of $(t - t_0)$. If the table interval is h, the next table values of x and y are obtained by the formulas

$$\left.\begin{aligned} x(t + h) &= x_0 + \int_{t_0}^{t+h} f(x, y, t)dt, \\ y(t + h) &= y_0 + \int_{t_0}^{t+h} \varphi(x, y, t)dt. \end{aligned}\right\} \quad (7)$$

But here we may, evidently, again calculate the integrals on the right by some quadrature-formula of the open type.

185. The problem is, thus, in principle solved. But in applying the method to individual cases a good many questions arise which are best illustrated by a numerical example. As calculations of this nature easily assume tedious proportions, and as quite a simple example serves the purpose of illustration equally well, we confine ourselves to a case where the reader will find it easy to go through all the calculations for himself as an exercise.

Let the given differential system be

$$\left.\begin{aligned} x' &= x - y + 2t - 1, \\ y' &= 2x - y + 3t + 1, \end{aligned}\right\} \quad (8)$$

and let the problem be to calculate a table of the integral which, for $t = 0$, assumes the values $x = 1, y = 0$. We must then calculate x and y by the equations

$$\left.\begin{aligned} x &= 1 + \int_0^t x'dt, \\ y &= \int_0^t y'dt, \end{aligned}\right\} \quad (9)$$

x' and y' having the significance (8). We assume that the values of x and y corresponding to $t = 0.0, 0.1, 0.2, 0.3, 0.4, 0.5$ (the framed numbers in the table below) have been given as initial values.

t	x	x'	$x^{(7)}$	y	y'	$y^{(7)}$
0.0	1.000000	0.000000	−1.00	0.000000	3.000000	−2.00
0.1	0.994837	−0.104830	−0.90	0.299667	2.990007	−1.99
0.2	0.978736	−0.218603	−0.78	0.597339	2.960133	−1.96
0.3	0.950856	−0.340184	−0.66	0.891040	2.910672	−1.91
0.4	0.910479	−0.468358	−0.53	1.178837	2.842121	−1.84
0.5	0.857007	−0.601844	−0.40	1.458851	2.755163	−1.76
0.6	0.789978	−0.739306	−0.26	1.729284	2.650672	−1.65
0.7	0.709060	−0.879376	−0.12	1.988436	2.529684	−1.53
0.8	0.614062	−1.020649	+0.02	2.234711	2.393413	−1.39
0.9	0.504937	−1.161718	+0.16	2.466655	2.243219	−1.24
1.0	0.381772			2.682941		

We may now, by (8), calculate the values of x' and y' corresponding to these arguments, and insert them in their places in the table.

The next question is the choice of quadrature-formula. In order to select the most suitable one, we want to know a few of the first derivatives of x and y. These may, even in the case of complicated forms of f and φ, be obtained by successive differentiations of (5), eliminating after each differentiation, by means of (5), the x' and y' introduced by the differentiation. In the particular case (8) under consideration we find

$$\left.\begin{aligned} x^{(7)} &= y - x - 2t \\ y^{(7)} &= y - 2x - 3t \end{aligned}\right\} \quad (10)$$

whence it follows that the five-term formula of the open type is sufficiently accurate for our purpose. As the interval for t is 0.1, we must first transform §16 (17) so that the interval of integration becomes 0.6. We need only multiply the remainder-term by $(0.6)^7$, and the principal terms by 0.6, after which the formula may be written

$$\int_t^{t+0.6} F(x)dx = 0.78F_0 + 0.33F_{\pm 2} - 0.42F_{\pm 1} + 0.0^731F^{(6)}(\xi). \quad (11)$$

It is seen, that as long as the numerical values of $x^{(6)}$ and $y^{(6)}$ do not exceed 16, the error due to the quadrature-formula (leaving the forcing-errors aside) cannot exceed one half unit of the sixth decimal A similar examination of the three-term formula shows, on the other hand, that this formula does not secure the accuracy of the sixth decimal, unless the interval is diminished.

We now proceed as follows. By means of the values of x' and y' already calculated we compute, by (11),

$$\left.\begin{aligned} x(0.6) &= 1 + \int_0^{0.6} x'\,dt, \\ y(0.6) &= \int_0^{0.6} y'\,dt, \end{aligned}\right\}$$

whereafter $x'(0.6)$ and $y'(0.6)$ are calculated by (8).

Next, we calculate, in the same way,

$$\left.\begin{aligned} x(0.7) &= x(0.1) + \int_{0.1}^{0.7} x'\,dt, \\ y(0.7) &= y(0.1) + \int_{0.1}^{0.7} y'\,dt, \end{aligned}\right\}$$

whereafter $x'(0.7)$ and $y'(0.7)$ are obtained by (8). In this way we may continue; we recommend to the reader to re-calculate a few of the values shown in the table, whereafter he will be perfectly familiar with the method.

The columns headed $x^{(7)}$ and $y^{(7)}$ have been added in order to control the remainder-term during the progress of the calculation. For this purpose it suffices to calculate these functions which are obtained by (10), to two places.

186. We must here mention the difficulty arising from the fact that the remainder-term depends on an argument ξ which may be any number between the limits of integration and, therefore, need not be one of the values for which $x^{(7)}$ and $y^{(7)}$ are known. It is true that we may regard the known values of $x^{(7)}$ and $y^{(7)}$ as *random samples* of the sixth differential coefficient; but we may also easily attain absolute certainty.

Let us assume that the functions f and φ, as is nearly always the case in practice, are *bounded in the neighbourhood of* (x_0, y_0, t_0). Or, to put it more precisely: f and φ must be of such a nature that if x, y and t were three *independent* variables, varying within a small region defined by the inequalities $|x - x_0| \leq k$, $|y - y_0| \leq k$, $|t - t_0| \leq h$, we have constantly $|f| \leq M$ and $|\varphi| \leq M$ where M is a constant, depending generally on x_0, y_0, t_0, k and h.

If x, y and t are not mutually independent but connected by the equations (6), the above inequalities still hold. Now it follows from (6) that it suffices to choose $|t - t_0|$ *sufficiently small* in order to ensure that $|x - x_0| \leq k$, $|y - y_0| \leq k$. If *also* $|t - t_0| \leq h$, we have $|f| \leq M$, $|\varphi| \leq M$, so that by (6)

$$\left.\begin{aligned} |x - x_0| &\leq |t - t_0| M, \\ |y - y_0| &\leq |t - t_0| M. \end{aligned}\right\} \quad (12)$$

If, now, we let $|t - t_0|$ increase but without exceeding h, it follows from (12) that, as long as only $|t - t_0| \leq \frac{k}{M}$, we have still $|x - x_0| \leq k$, $|y - y_0| \leq k$.

We have thus proved the following theorem:

If the variation of t does not exceed the smaller of the numbers h and $\frac{k}{M}$, *the variations of x and y cannot exceed k.*

This theorem can evidently be extended to any number of equations.

It is easily applied to the case under consideration. We have here

$$\left.\begin{aligned} f &= x' = x - y + 2t - 1, \\ \varphi &= y' = 2x - y + 3t + 1, \end{aligned}\right\}$$

consequently, if we give x, y and t the increments Δx, Δy, and Δt,

$$\left.\begin{aligned} |f| &\leq |x'| + |\Delta x| + |\Delta y| + 2|\Delta t|, \\ |\varphi| &\leq |y'| + 2|\Delta x| + |\Delta y| + 3|\Delta t|. \end{aligned}\right\}$$

We have therefore at any point of the section of the table calculated above

$$M < 3 + 3k + 3h.$$

Now the variation of t in the table interval is $h = 0.1$. If we take $k = 0.5$ we have $M < 4.8$, consequently $h < \frac{k}{M}$. It follows that x and y cannot vary by more than 0.5, if the variation of t does not exceed 0.1. But from this we may, by means of (10), conclude that $x^{(7)}$ and $y^{(7)}$ cannot vary by more than 1.8 within the table interval, so that we have been justified in neglecting the remainder-term.

An examination of this kind should, of course, not be made at each step of the calculation but, as we have done, for a suitable section of the completed table at the time.

187. A check on the calculation may, for instance, be obtained by the five-term formula of the closed type which, after transforming the interval, may be written

$$\int_t^{t+0.4} F(x)\,dx = \frac{24F_0 + 64F_{\pm 1} + 14F_{\pm 2}}{450} - 0.0^9 86\,F^{(6)}(\xi). \quad (13)$$

If this check is applied step by step, we also guard against losing in accuracy, a point of considerable importance in calculations of this nature where the work is much increased by carrying superfluous decimals. Suppose, to take an extreme case, that F_{-2}, F_{-1}, F_0 and F_1 have 6 correct decimals, but F_2 only 5. In this case, the error in x and y calculated by (13), inasmuch as it is due to F_2, is smaller than $0.0^5 5 \times \frac{14}{450} < 0.0^6 16$, so that we obtain the correct values of x and y, by which F_2 can be corrected.

188. In the course of a lengthy calculation it often happens that an interval which was suitable at the beginning of the calculation, later on proves either too large, or smaller than necessary. In such cases we may either choose a more suitable quadrature-formula or, as is often preferable, change the interval. Transition to a larger interval does not present any difficulties, but in reducing the interval it becomes necessary first to procure the necessary initial values, e.g. by interpolation in the section of the table already calculated.

189. In the numerical example we chose for illustration, the calculation of f and φ for given values of x, y and t was a very small matter; but in the cases where numerical integration is of real practical use, this is the heaviest part of the work and, unfortunately, little can be done in order to facilitate this part of the calculation. Also the remainder-term can cause difficulties, if f and φ are not very simple; but we may often profit by the fact that we need not possess the exact expression of $F^{(2k)}(\xi)$ but only a suitable upper limit to this function. What liberties may be taken in this respect depends, however, on the particular nature of f and φ, and it is hardly possible to give general instructions on this point.

190. We need not point out that the differential system (8), used above as an illustration, is easily completely integrated, and that the special integral we have traced by numerical integration is

$$\left.\begin{aligned} x &= \cos t + \sin t - t, \\ y &= 2\sin t + t. \end{aligned}\right\} \quad (14)$$

On comparing the values calculated above with the values resulting from (14), it is seen that the error nowhere exceeds one unit of the sixth decimal, as might be expected.

191. Another method of numerical integration of differential equations may, as mentioned in No. 131, be derived from the Gaussian summation-formulas. The values of $F(x)$ occurring in the corrective terms at the upper limit must, in that case, be obtained by extrapolation. We shall not go into details.[1]

[1] See a recent paper by E. J. Nyström: Über die numerische Integration von Differentialgleichungen, Acta Societatis Scientiarum Fennicæ, Helsingfors, 1925.

§18. The Calculus of Symbols

192. We have already shown in §2 that certain symbols, denoting operations, may to some extent be treated as if they were numbers. More generally, let Ω denote an operation which possesses the *distributive, commutative,* and *associative* properties. Symbols, possessing all these three properties, we shall call *omega-symbols.* It is then seen, as in §2, that it is allowable to form polynomials in Ω and combine such polynomials with each other, as long as we confine ourselves to the processes of addition, subtraction and multiplication. In this section we shall, without losing sight of mathematical rigour, extend these rules by introducing *division* with an omega-symbol, and by showing that these symbols may sometimes, although with essential reservations, be expanded in *infinite series.*

193. Division by Ω, or multiplication by Ω^{-1}, is an operation of such a nature that, on multiplying the result by Ω, we obtain the original function, that is

$$\Omega\,\Omega^{-1} = 1. \tag{1}$$

The symbol Ω^{-1} *is not always an omega-symbol;* for, although it evidently possesses the distributive property, it is not always commutative. Let us, in fact, assume that we have found a function $\varphi(x)$, such that $\Omega\varphi(x) = f(x)$, $f(x)$ being a given function. Then, according to the definition of division,

$$\Omega^{-1}f(x) = \varphi(x) + \psi(x), \tag{2}$$

$\psi(x)$ denoting any function for which $\Omega\psi(x) = 0$, or $\psi(x) = \Omega^{-1}0$. If now, in (2), we insert $f(x) = \Omega\varphi(x)$, we get

$$\Omega^{-1}\,\Omega\varphi(x) = \varphi(x) + \psi(x),$$

so that only in the case of $\psi(x) = 0$ do we have $\Omega^{-1}\,\Omega = 1$ and, consequently, $\Omega^{-1}\,\Omega = \Omega\,\Omega^{-1}$.

We have, for instance,

$$\Delta^{-1}f(x) = \varphi(x) + \psi(x),$$

$\varphi(x)$ being a function such that $\Delta\varphi(x) = f(x)$, while $\psi(x)$ need only satisfy the condition $\Delta\psi(x) = 0$ and, therefore, may be *any periodic function with the period* 1.

It is seen that Δ^{-1} is, in reality, a summation-symbol, and that $\psi(x)$ vanishes in the case of summation between definite limits. This function is, therefore, an analogue of the arbitrary constant implied in an indefinite integral. In the frequent cases where the result of the summation is only of interest to us for integral values of the argument, we may replace $\psi(x)$ by a constant; and for this reason that function is often termed somewhat illogically *a periodic constant.*

Similar remarks may be made with respect to the symbols ∇^{-1} and δ^{-1}; as regards D^{-1}, it is immediately seen that this symbol is identical with a sign of integration.

194. We shall now consider the special omega-symbol $\Omega = E^a - c$, a and c being constants. If $c = 0$, $\Omega^{-1} = E^{-a}$ is an omega-symbol, as follows from the remarks in No. 6; it is also seen directly that if $E^a\psi(x) = 0$, we must have $\psi(x) = 0$. But if $c \neq 0$, it becomes necessary to examine whether the equation

$$(E^a - c)\,\psi(x) = 0 \tag{3}$$

has a solution that differs from zero.

If, in this equation, we put

$$\psi(x) = c^{\frac{x}{a}}\,F(x),$$

we get

$$(E^a - c)\,c^{\frac{x}{a}}\,F(x) = c^{\frac{x}{a}+1}[F(x + a) - F(x)] = 0,$$

and this equation is satisfied, if for $F(x)$ we take a periodic function with the period a, and not otherwise. The solution of (3) can therefore be written

$$\psi(x) = c^{\frac{x}{a}}\,\psi_a(x), \tag{4}$$

$\psi_a(x)$ being a periodic function with the period a, but otherwise arbitrary.

The symbol $(E^a - c)^{-1}$ is consequently, for $c \neq 0$, never an omega-symbol, but in applying it, we introduce a function of the form (4).

195. If, more generally, we consider the omega-symbol $\Omega = (E^a - c)^n$, where $c \neq 0$, while n is a positive integer, we must examine the equation

$$(E^a - c)^n\,\psi(x) = 0. \tag{5}$$

If we put $\psi(x) = c^{\frac{x}{a}} F(x)$, it may be shown by induction that

$$(E^a - c)^n c^{\frac{x}{a}} F(x) = c^{\frac{x}{a}+n} \underset{a}{\Delta}^n F(x) = 0, \tag{6}$$

so that we have, for the determination of $F(x)$, the equation

$$\underset{a}{\Delta}^n F(x) = 0.$$

But from this we obtain by successive multiplications by $\underset{a}{\Delta}^{-1} = (E^a - 1)^{-1}$, denoting by $c_1, c_2, c_3, \ldots$ periodic constants with the period a, and putting for the moment $x^{(\nu)} \equiv x(x - a) \ldots (x - \nu a + a)$,

$$F(x) = \frac{c_1}{(n-1)!} x^{(n-1)} + \frac{c_2}{(n-2)!} x^{(n-2)} + \ldots + c_n,$$

so that $F(x)$ is a polynomial of degree $(n - 1)$ in x with coefficients which are arbitrary periodic constants with the period a.

196. By the same simple method we may examine omega-symbols consisting of a finite number of factors of the form $(E^a - c)^n$, and are thus enabled to deal with certain simple *difference-equations*,[1] that is, equations containing an unknown function and its differences up to a certain order, or—what comes to the same thing—a number of unknown values of a function, corresponding to the equidistant arguments $x, x + h, x + 2h, \ldots x + mh$. We content ourselves with illustrating the method by means of a few simple examples.[2]

Let the given equation be

$$f(x + 2) - 4f(x + 1) + 4f(x) = 3^x x$$

or—which comes to the same—

$$f(x) - 2\Delta f(x) + \Delta^2 f(x) = 3^x x;$$

an equation which may be written

$$(E - 2)^2 f(x) = 3^x x. \tag{7}$$

[1] The subject as a whole falls outside the scope of this book. We refer the reader to N. E. Nörlund: Differenzenrechnung, Berlin 1924, being the most modern exposition of the subject.

[2] G. Boole: The Calculus of Finite Differences, third ed., chapter XI.

It is easy to see that this equation has a particular solution of the form

$$f_1(x) = 3^x(k_1 x + k_2);$$

for, inserting this expression in (7), we may determine the constants and find $k_1 = 1$, $k_2 = -6$. If, therefore, we put

$$f(x) = 3^x(x - 6) + \psi(x),$$

we get

$$(E - 2)^2\psi(x) = 0, \quad \psi(x) = (E - 2)^{-2}0 = 2^x(c_1 + c_2 x),$$

and consequently, c_1 and c_2 being periodic constants with the period 1,

$$f(x) = 3^x(x - 6) + 2^x(c_1 + c_2 x). \tag{8}$$

197. In a similar way we may solve the difference-equation

$$\Delta^2 f(x) - 3\Delta f(x) + 2f(x) = a^x$$

or

$$(E - 3)(E - 2)f(x) = a^x. \tag{9}$$

In this case, there is generally a particular solution of the form $f_1(x) = ka^x$, as we find, by insertion, $k = \dfrac{1}{(a - 2)(a - 3)}$, provided only that a is different from 2 and 3.

For the determination of $\psi(x)$ we have now

$$(E - 3)(E - 2)\psi(x) = 0,$$

whence

$$(E - 2)\psi(x) = (E - 3)^{-1}0$$

or

$$(E - 2)\psi(x) = 3^x c_1.$$

But this equation has the particular solution $\psi(x) = 3^x c_1$, so that its complete solution is

$$\begin{aligned}\psi(x) &= 3^x c_1 + (E - 2)^{-1}0 \\ &= 3^x c_1 + 2^x c_2.\end{aligned}$$

Therefore the complete solution of (9) is

$$f(x) = \frac{a^x}{(a-2)(a-3)} + 3^x c_1 + 2^x c_2, \tag{10}$$

provided that a is different from 2 and 3.

If $a = 3$, there is a particular solution of the form $f_1(x) = kx.3^x$. We find $k = \frac{1}{3}$ and

$$f(x) = 3^{x-1}x + 3^x c_1 + 2^x c_2.$$

If $a = 2$, we find $f_1(x) = kx.\, 2^x$, $k = -\frac{1}{2}$ and

$$f(x) = -2^{x-1}x + 3^x c_1 + 2^x c_2.$$

198. In solving difference-equations by the calculus of symbols, we are often confronted by imaginary expressions which, however, are no cause of real difficulties, as our symbols evidently combine with complex numbers according to the same rules as with real numbers.

Thus, from the difference-equation

$$f(x+2) + f(x) = 0$$

or

$$(E+i)(E-i)f(x) = 0$$

we obtain, proceeding in the usual way, the solution

$$\begin{aligned} f(x) &= i^x c_1 + (-i)^x c_2 \\ &= c_1 e^{\frac{i\pi x}{2}} + c_2 e^{-\frac{i\pi x}{2}} \\ &= a \cos \frac{\pi x}{2} + b \sin \frac{\pi x}{2}, \end{aligned}$$

a and b being, like c_1 and c_2, periodic constants with the period 1.

Treating the same example by the method of No. 195, we write the equation

$$(E^2 + 1)f(x) = 0$$

and find

$$f(x) = K(-1)^{\frac{x}{2}} = (A + iB)\left(\cos \frac{\pi x}{2} + i \sin \frac{\pi x}{2}\right);$$

but as the real and imaginary parts of $f(x)$ must separately satisfy the difference-equation, we evidently have the same solution as before.

199. We are now in a position to solve *the linear homogeneous difference-equation with constant coefficients, or*

$$a_0 f(x) + a_1 f(x+1) + \dots + a_\kappa f(x+\kappa) = 0,$$

an equation which has many applications for instance in the theory of probabilities.

We write the equation in symbolical form

$$(a_0 + a_1 E + \dots + a_\kappa E^\kappa) f(x) = 0.$$

Resolving the polynomial into factors and dividing by a_κ, we obtain the form

$$(E - \alpha_1)^{n_1}(E - \alpha_2)^{n_2} \dots (E - \alpha_\mu)^{n_\mu} f(x) = 0,$$

the numbers α_ν being all different, and the exponents n_ν being positive integers with the sum κ.

We find now, multiplying by $(E - \alpha_1)^{-n_1}$,

$$(E - \alpha_2)^{n_2} \dots (E - \alpha_\mu)^{n_\mu} f(x) = \alpha_1^x F_{n_1-1}(x),$$

$F_{n_1-1}(x)$ being a polynomial of degree $n_1 - 1$ with coefficients that are periodic constants with the period 1.

From this we obtain, multiplying by $(E - \alpha_2)^{-n_2}$,

$$(E - \alpha_3)^{n_3} \dots (E - \alpha_\mu)^{n_\mu} f(x) = \alpha_1^x \Phi_{n_1-1}(x) + \alpha_2^x F_{n_2-1}(x),$$

$\Phi_{n_1-1}(x)$ being a polynomial of *the same degree* as $F_{n_1-1}(x)$. We realize this by noticing that if we put (apart from the arbitrary function)

$$(E - \alpha_2)^{-1} \alpha_1^x F_{n_1-1}(x) = \alpha_1^x \Phi(x),$$

we have, for the determination of the coefficients in $\Phi(x)$, as is seen on multiplying by $(E - \alpha_2)$,

$$F_{n_1-1}(x) = \alpha_1 \Phi(x+1) - \alpha_2 \Phi(x),$$

so that the coefficient of x^{n_1-1} cannot vanish, as $\alpha_1 \neq \alpha_2$. The degree of the polynomial is, therefore, not changed, if we repeat the operation $(E - \alpha_2)^{-1}$ any number of times, say, n_2 times.

If we continue in the same way, we obtain finally the complete solution of the linear homogeneous difference-equation with constant coefficients, in the form

$$f(x) = \sum_{\nu=1}^{\mu} \alpha_\nu^x F_{n_\nu - 1}(x).$$

200. We now turn our attention to a particular class of omega-symbols which will be called *theta-symbols*. A theta-symbol, Θ, must, like all other omega-symbols, possess the distributive, commutative and associative properties; but besides these, it must have one characteristic property: if $P(x)$ is a polynomial, $\Theta P(x)$ must be a polynomial of *lower degree*, while $\Theta c = 0$, if c is a constant. The symbols Δ, ∇, δ, D and the positive integral powers of these symbols are, therefore, examples of theta-symbols, while E and $\square$ are only omega-symbols, but not theta-symbols.

Now, if the object of the operations is a polynomial of degree n in x, we evidently obtain the same result whether, on this polynomial, we perform an operation which is a polynomial of degree n in Θ, or perform the operation when to the polynomial in Θ we have added terms of higher degree than the nth, as these cannot influence the result. We may, therefore, form infinite power-series in Θ, according to the same rules as if Θ were a sufficiently small number, and perform calculations with such series in the ordinary way, provided only that the operations defined thereby are only applied to *polynomials* in x.[1]

While, for instance, Log $(1 + \Delta)$ has not in itself any meaning, the expression

$$\Delta - \frac{\Delta^2}{2} + \frac{\Delta^3}{3} - \frac{\Delta^4}{4} + \cdots$$

has a well defined meaning, if applied to a polynomial, in which case only a finite number of terms of the expansion are used; and nothing prevents us from putting

$$\text{Log}\,(1 + \Delta) = \Delta - \frac{\Delta^2}{2} + \frac{\Delta^3}{3} - \frac{\Delta^4}{4} + \cdots,$$

[1] J. L. W. V. Jensen: Sur une identité d'Abel et sur d'autres formules analogues. Acta Mathematica, vol. 26, p. 314.

provided the operation Log $(1 + \Delta)$, thus defined, is only applied to polynomials.

Similarly, by $\dfrac{e^\delta}{1-\delta}$ we mean the series obtained by expanding this expression in powers of δ; but this series being infinite, we only allow its application to polynomials in x. Under these circumstances it is evidently the product of the operations e^δ and $\dfrac{1}{1-\delta}$ if, by these, we understand the operations

$$1 + \frac{\delta}{1!} + \frac{\delta^2}{2!} + \frac{\delta^3}{3!} + \dots$$

and

$$1 + \delta + \delta^2 + \delta^3 + \dots,$$

and only apply them to polynomials.

On the other hand, we cannot attach any meaning whatever to a symbol such as, for instance, Log Δ, as it is not possible to expand this expression, even formally, in powers of Δ or any other theta-symbol. In our calculations with symbols we must, therefore, carefully avoid using expressions which cannot be expanded in powers of some theta-symbol.

201. We call the attention of the reader, *once and for all,* to the fact that, wherever we introduce *infinite* power-series in theta-symbols, either directly, or indirectly through functional expressions, the tacit assumption is *that the object of these operations is a polynomial.* It is true that the field of application of this calculus thus becomes very restricted; nevertheless, it is of considerable use, because the form of an interpolation- or summation-formula, that is, its terms with the exception of the remainder-term, do not depend on whether the function is a polynomial or not. The calculus of symbols is therefore, as we shall presently see, an easy and reliable instrument, always at hand, for reproducing a great number of the results we have derived by other methods.

The calculus of symbols has also a stimulating effect on the science of finite differences, as very often a theorem has first, by symbolical methods, been proved for polynomials, and afterwards been extended to more general classes of functions.[1]

202. Δ being a theta-symbol, it is immediately seen that the displacement-symbol E^a may be developed in powers of Δ, as

$$E^a = (1 + \Delta)^a = \sum_0^\infty \binom{a}{\nu} \Delta^\nu. \tag{11}$$

If $P(x)$ is a polynomial of degree n, this relation only expresses that

$$P(x + a) = \sum_0^n \frac{a^{(\nu)}}{\nu!} \Delta^\nu P(x),$$

as was already proved in No. 14.

Now Log $(1 + \Delta)$, as is seen on expanding in powers of Δ, is also a theta-symbol, and we therefore find, by expanding E^a in powers of Log $(1 + \Delta)$,

$$E^a = e^{a \operatorname{Log}(1+\Delta)} = \sum_0^\infty \frac{a^\nu}{\nu!} \operatorname{Log}^\nu(1 + \Delta).$$

But this relation is nothing but the expansion of $P(x + a)$ in powers of a, or

$$P(x + a) = \sum_0^n \frac{a^\nu}{\nu!} D^\nu P(x),$$

whence follows that

$$D = \operatorname{Log}(1 + \Delta) = \operatorname{Log} E, \tag{12}$$

so that we obtain Taylor's formula in the form

$$E^a = e^{aD} = \sum_0^\infty \frac{a^\nu}{\nu!} D^\nu. \tag{13}$$

[1] The calculus of symbols has been extended to other classes of functions than polynomials by S. Pincherle: Le operazioni distributive, Bologna, 1901; Funktionaloperationen und -gleichungen (Encykl. der math. Wissenschaften, II A 11).

The interpolation-formula with ascending differences is, of course, obtained by expanding E^a in powers of ∇; we find

$$E^a = (1 - \nabla)^{-a} = \sum_0^\infty \frac{a^{(-\nu)}}{\nu!} \nabla^\nu. \tag{14}$$

From (12) results immediately

$$D = \operatorname{Log}(1 + \Delta) = \sum_1^\infty (-1)^{\nu+1} \frac{\Delta^\nu}{\nu}; \tag{15}$$

we have also

$$D = -\operatorname{Log}(1 - \nabla) = \sum_1^\infty \frac{\nabla^\nu}{\nu}. \tag{16}$$

203. The formulas with central differences are not quite so obvious. We find from §2 (13)

$$\left.\begin{aligned} E^{\frac{1}{2}} &= \frac{\delta}{2} + \sqrt{1 + \frac{\delta^2}{4}} \\ &= 1 + \frac{\delta}{2} + \sum_{\nu=1}^{\infty} \binom{\frac{1}{2}}{\nu} \left(\frac{\delta}{2}\right)^{2\nu}; \end{aligned}\right\} \tag{17}$$

the sign of the square-root has been determined by applying the operation $E^{\frac{1}{2}}$ to a constant.

But from (17) follows

$$E^a = \left(\frac{\delta}{2} + \sqrt{1 + \frac{\delta^2}{4}}\right)^{2a}, \tag{18}$$

and expansion of the right-hand side in powers of δ must result in the interpolation-formula with central differences.

If we want the general term of this expansion, we can do no better than refer back to No. 14, where it was proved simply enough that

$$P(x + a) = \sum_0^n \frac{a^{[\nu]}}{\nu!} \delta^\nu P(x)$$

or in symbolical form

$$E^a = \sum_0^\infty \frac{a^{[\nu]}}{\nu!}\delta^\nu, \tag{19}$$

being, therefore, the expansion of (18).

From (12) and (18) we find for $a = 1$

$$D = 2\text{Log}\left(\frac{\delta}{2} + \sqrt{1 + \frac{\delta^2}{4}}\right) = \int_0^\delta \left(1 + \frac{x^2}{4}\right)^{-\frac{1}{2}} dx. \tag{20}$$

Expanding the integrand, and integrating from 0 to δ, we get

$$D = \sum_0^\infty (-1)^\nu \frac{[1.3 \;.\;.\;.\; (2\nu - 1)]^2}{2^{2\nu}(2\nu + 1)!}\delta^{2\nu+1}. \tag{21}$$

204. The symbol $\Box^{-1}$ can evidently be expanded in powers of δ

$$\Box^{-1} = \left(1 + \frac{\delta^2}{4}\right)^{-\frac{1}{2}} = \sum_{\nu=0}^\infty \binom{-\frac{1}{2}}{\nu}\left(\frac{\delta}{2}\right)^{2\nu} = \frac{dD}{d\delta}, \tag{22}$$

and it is seen that $\Box^{-1}$, as well as $\Box$ itself, is an omega-symbol.

If now we put

$$\delta_{2a} = E^a - E^{-a}, \qquad 2\Box_{2a} = E^a + E^{-a}, \tag{23}$$

we have evidently

$$E^a = \Box_{2a} + \tfrac{1}{2}\delta_{2a}, \tag{24}$$

and find from (23), by (19),

$$\Box_{2a} = \sum_{\nu=0}^\infty \frac{a^{[2\nu]}}{(2\nu)!}\delta^{2\nu}. \tag{25}$$

Observing that (22) may be written

$$\frac{d\,\text{Log}\,E}{d\delta} = \Box^{-1}$$

whence

$$\frac{dE}{d\delta} = E\Box^{-1},$$

we obtain from (23)

$$\frac{d\Box_{2a}}{d\delta} = \tfrac{1}{2}a\delta_{2a}E^{-1}\frac{dE}{d\delta} = \tfrac{1}{2}\, a\delta_{2a}\,\Box^{-1},$$

so that, on differentiating (25) with respect to δ, we find

$$\frac{1}{2}\,\delta_{2a} = \sum_{\nu=0}^{\infty}\frac{a^{[2\nu+2]-1}}{(2\nu+1)\,!}\,\Box\,\delta^{2\nu+1}. \tag{26}$$

If finally we add together this formula and (25), we get

$$E^{a} = \sum_{\nu=0}^{\infty}\left[\frac{a^{[2\nu]}}{(2\nu)\,!}\,\delta^{2\nu} + \frac{a^{[2\nu+2]-1}}{(2\nu+1)\,!}\,\Box\,\delta^{2\nu+1}\right] \tag{27}$$

or Stirling's interpolation-formula.

If, from this, we form

$$\lim_{a\to 0}\frac{E^{a}-1}{a} = D,$$

we find

$$D = \sum_{0}^{\infty}(-1)^{\nu}\frac{(\nu\,!)^{2}}{(2\nu+1)\,!}\,\Box\,\delta^{2\nu+1}. \tag{28}$$

Bessel's formula may be derived in much the same way. We have, by (19),

$$\frac{1}{2}\,\delta_{2a} = \sum_{\nu=0}^{\infty}\frac{a^{[2\nu+1]}}{(2\nu+1)\,!}\,\delta^{2\nu+1} \tag{29}$$

whence, by differentiation with respect to δ,

$$\Box_{2a} = \sum_{\nu=0}^{\infty}\frac{a^{[2\nu+1]-1}}{(2\nu)\,!}\,\Box\,\delta^{2\nu} \tag{30}$$

and finally, by adding (29) and (30),

$$E^{a} = \sum_{\nu=0}^{\infty}\left[\frac{a^{[2\nu+1]-1}}{(2\nu)\,!}\,\Box\,\delta^{2\nu} + \frac{a^{[2\nu+1]}}{(2\nu+1)\,!}\,\delta^{2\nu+1}\right]. \tag{31}$$

Everett's first formula is obtained if, in the identity

$$E^x = \frac{\delta_{2x}}{2\Box\delta}E + \frac{\delta_{2(1-x)}}{2\Box\delta},$$

we introduce the expressions for δ_{2x} and $\delta_{2(1-x)}$ from (26).

In order to derive Everett's second formula, we may, in the identity

$$E^{x+\frac{1}{2}} = E + \frac{\Box_{2x} - \Box}{\Box\delta}E^{\frac{3}{2}} - \frac{\Box_{2(1-x)} - \Box}{\Box\delta}E^{\frac{1}{2}},$$

introduce $\Box_{2x}$ and $\Box_{2(1-x)}$, by (30).

205. As regards the summation problems, we have already said that, in summing between definite limits, the periodic constant vanishes. We may, for instance, put

$$1 + E + E^2 + \ldots + E^{n-1} = \frac{E^n - 1}{E - 1}, \tag{32}$$

provided the right-hand side is interpreted as $(E^n - 1)\Delta^{-1}$; for both sides, if developed in powers of Δ, lead to the same series. The reason for this is that the ambiguity attached to the symbol Δ^{-1} is removed by the factor $(E^n \dot{-} 1)$, cancelling the periodic constant. In symbolical formulas where Δ^{-1} enters, it is customary to leave out the factor $(E^n - 1)$ which must be mentally added, and we shall follow this convenient practice.

If, therefore, we want to derive, say, Laplace's summation formula, the problem is simply to expand D^{-1} in powers of Δ, the result being

$$D^{-1} = \frac{1}{\operatorname{Log}(1 + \Delta)} = \Delta^{-1} + \sum_1^\infty L_\nu \Delta^{\nu-1}. \tag{33}$$

206. In order to obtain the first Gaussian summation-formula, we must expand D^{-1} in powers of δ, that is

$$D^{-1} = \frac{1}{2\operatorname{Log}\left(\frac{\delta}{2} + \sqrt{1 + \frac{\delta^2}{4}}\right)} = \delta^{-1} + \sum_{\nu=1}^\infty K_{2\nu}\delta^{2\nu-1}. \tag{34}$$

The factor left out on both sides is, in this case $(E^{m-\frac{1}{2}} - E^{-\frac{1}{2}})$; if this factor is added, the first term on the right means the sum from 0 to $m - 1$, as $E^{-\frac{1}{2}}\delta^{-1} = \Delta^{-1}$.

In order to obtain the coefficients $K_{2\nu}$, we multiply by D on both sides of (34), and find, by (21),

$$1 = (\delta^{-1} + K_2\delta + K_4\delta^3 + \ldots)\left(\delta - \frac{1^2}{3!\,2^2}\delta^3 + \frac{(1.3)^2}{5!\,2^4}\delta^5 - \frac{(1.3.5)^2}{7!\,2^6}\delta^7 + \ldots\right).$$

If, now, we examine the coefficient of $\delta^{2\nu}$ on both sides, we find the recurrence formula

$$\left.\begin{aligned} K_{2\nu} &= \frac{1^2}{3!\,2^2}K_{2\nu-2} - \frac{(1.3)^2}{5!\,2^4}K_{2\nu-4} + \ldots \\ &+ (-1)^\nu \frac{[1.3\ldots(2\nu-3)]^2}{(2\nu-1)!\,2^{2\nu-2}}K_2 + (-1)^{\nu+1}\frac{[1.3\ldots(2\nu-1)]^2}{(2\nu+1)!\,2^{2\nu}}, \end{aligned}\right\} \quad (35)$$

leading, of course, to the same values as the method employed in No. 122.

207. The second Gaussian formula is obtained, if (34) is multiplied by

$$1 = \square\left(1 + \frac{\delta^2}{4}\right)^{-\frac{1}{2}},$$

whence

$$D^{-1} = \frac{\square}{\sqrt{1 + \frac{\delta^2}{4}}\cdot 2\,\mathrm{Log}\left(\frac{\delta}{2} + \sqrt{1 + \frac{\delta^2}{4}}\right)} = \square\,\delta^{-1} + \sum_{\nu=1}^{\infty} M_{2\nu}\square\,\delta^{2\nu-1}. \quad (36)$$

The factor left out is $(E^m - 1)$. For the coefficients a recurrence formula may be found. For if we write (36) in the form

$$\left(1 + \frac{\delta^2}{4}\right)^{-\frac{1}{2}} = \left(\delta^{-1} + \sum_{\nu=1}^{\infty} M_{2\nu}\delta^{2\nu-1}\right)\cdot 2\,\mathrm{Log}\left(\frac{\delta}{2} + \sqrt{1 + \frac{\delta^2}{4}}\right)$$

and examine the coefficient of $\delta^{2\nu}$ on both sides, expanding the right-hand side by (20) and (21), we get

$$\left.\begin{aligned} M_{2\nu} &= \frac{1^2}{3!\,2^2}M_{2\nu-2} - \frac{(1.3)^2}{5!\,2^4}M_{2\nu-4} + \ldots \\ &+ (-1)^\nu \frac{[1.3\ldots(2\nu-3)]^2}{(2\nu-1)!\,2^{2\nu-2}}M_2 + (-1)^\nu \frac{[1.3\ldots(2\nu-1)]^2}{(2\nu-1)!\,(2\nu+1)2^{2\nu}}. \end{aligned}\right\} \quad (37)$$

The values found by this formula are identical with those calculated in No. 124.

208. The first Eulerian summation-formula is simply the expansion of Δ^{-1} in powers of D, or

$$\Delta^{-1} = \frac{1}{e^D - 1} = \sum_0^\infty \frac{B_\nu}{\nu!} D^{\nu-1}, \tag{38}$$

while the second Eulerian formula is the expansion of δ^{-1} in powers of D, or

$$\delta^{-1} = \frac{1}{e^{\frac{D}{2}} - e^{-\frac{D}{2}}} = \sum_0^\infty \frac{B_\nu(\frac{1}{2})}{\nu!} D^{\nu-1}. \tag{39}$$

209. We put together, for reference, the first few terms of these summation-formulas, as they are used in practice.

$$\left.\begin{aligned}
D^{-1} &= \Delta^{-1} + \frac{1}{2} - \frac{1}{12}\Delta + \frac{1}{24}\Delta^2 - \frac{19}{720}\Delta^3 + \frac{3}{160}\Delta^4 - \frac{863}{60480}\Delta^5 + \ldots \\
D^{-1} &= \delta^{-1} + \frac{1}{24}\delta - \frac{17}{5760}\delta^3 + \frac{367}{967680}\delta^5 - \ldots \\
D^{-1} &= \square\,\delta^{-1} - \frac{1}{12}\square\,\delta + \frac{11}{720}\square\,\delta^3 - \frac{191}{60480}\square\,\delta^5 + \ldots \\
D^{-1} &= \Delta^{-1} + \frac{1}{2} - \frac{1}{12}D + \frac{1}{720}D^3 - \frac{1}{30240}D^5 + \ldots \\
D^{-1} &= \delta^{-1} + \frac{1}{24}D - \frac{7}{5760}D^3 + \frac{31}{967680}D^5 - \ldots
\end{aligned}\right\} \tag{40}$$

210. Lubbock's formula may also be obtained by the calculus of symbols. In this case the problem is to develop Δ^{-1} in powers of $\Delta_h = E^h - 1$. As

$$\Delta = (1 + \Delta_h)^{\frac{1}{h}} - 1,$$

we find

$$\Delta^{-1} = \frac{1}{(1 + \Delta_h)^{\frac{1}{h}} - 1} = h\,\Delta_h^{-1} + \sum_1^\infty A_\nu \Delta_h^{\nu-1}, \tag{41}$$

the factor left out on both sides being $(E^{hk} - 1)$.

211. In order to derive the second formula of Lubbock's type, we must expand δ^{-1} in powers of

$$\delta_h = E^{\frac{h}{2}} - E^{-\frac{h}{2}}.$$

We obtain from this relation

$$E = \left(\frac{1}{2}\delta_h + \sqrt{1 + \frac{1}{4}\delta_h^2}\right)^{\frac{2}{h}}, \tag{42}$$

so that

$$\delta = \left(\frac{1}{2}\delta_h + \sqrt{1 + \frac{1}{4}\delta_h^2}\right)^{\frac{1}{h}} - \left(\frac{1}{2}\delta_h + \sqrt{1 + \frac{1}{4}\delta_h^2}\right)^{-\frac{1}{h}}. \tag{43}$$

Now (18) and (19) show that the following expansion exists

$$\left(\frac{x}{2} + \sqrt{1 + \frac{x^2}{4}}\right)^c = \sum_0^\infty \frac{1}{\nu!}\left(\frac{c}{2}\right)^{[\nu]} x^\nu; \tag{44}$$

we therefore find, from (43),

$$\delta = \sum_{\nu=0}^\infty \frac{2}{(2\nu+1)!}\left(\frac{1}{2h}\right)^{[2\nu+1]} \delta_h^{2\nu+1} \tag{45}$$

and in a similar way

$$\square = \sum_{\nu=0}^\infty \frac{1}{(2\nu)!}\left(\frac{1}{2h}\right)^{[2\nu]} \delta_h^{2\nu}. \tag{46}$$

The required expansion has the form

$$\delta^{-1} = h\delta_h^{-1} + P_2\,\delta_h + P_4\,\delta_h^3 + P_6\,\delta_h^5 + \dots \tag{47}$$

If we multiply both sides by δ and make use of (45), we find

$$1 = \left[\frac{1}{h}\delta_h + \frac{2}{3!}\left(\frac{1}{2h}\right)^{[3]}\delta_h^3 + \frac{2}{5!}\left(\frac{1}{2h}\right)^{[5]}\delta_h^5 + \dots\right]\left(h\delta_h^{-1} + P_2\delta_h + P_4\delta_h^3 + \dots\right)$$

whence, on comparing powers of δ_h on both sides, the recurrence formula

$$\left.\begin{aligned} -P_{2\nu} &= \frac{P_{2\nu-2}}{3!}\left(\frac{1}{2h}\right)^{[3]-1} + \frac{P_{2\nu-4}}{5!}\left(\frac{1}{2h}\right)^{[5]-1} + \dots \\ &+ \frac{P_2}{(2\nu-1)!}\left(\frac{1}{2h}\right)^{[2\nu-1]-1} + \frac{h}{(2\nu+1)!}\left(\frac{1}{2h}\right)^{[2\nu+1]-1}. \end{aligned}\right\} \tag{48}$$

The coefficients $P_{2\nu}$ calculated by this formula are, of course, identical with those calculated in No. 158 by another method.

It appears from (48) that these coefficients have the form of $\frac{h^2 - 1}{h^{2\nu-1}}$ multiplied by a polynomial in h^2.

The factor left out in (47) is $(E^{hk-\frac{1}{2}} - E^{-\frac{1}{2}})$; if this is taken into account, the formula is seen to be identical with §15(11).

212. The third formula of Lubbock's type is obtained by developing $\Box\,\delta^{-1}$ in terms of $\Box_h\,\delta_h^{2\nu-1}$. For it is, to begin with, clear that §15(15) may be written in the symbolical form

$$\Delta^{-1} = h\Delta_h^{-1} + \frac{h-1}{2} + \sum_{\nu=1}^{\infty} Q_{2\nu}\Box_h\delta_h^{2\nu-1},$$

the factor left out being $(E^{hk} - 1)$; but this formula is identical with

$$\Delta^{-1} + \tfrac{1}{2} = h(\Delta_h^{-1} + \tfrac{1}{2}) + \sum_{\nu=1}^{\infty} Q_{2\nu}\Box_h\delta_h^{2\nu-1}$$

and therefore with

$$\Box\,\delta^{-1} = h\Box_h\,\delta_h^{-1} + \sum_{\nu=1}^{\infty} Q_{2\nu}\Box_h\delta_h^{2\nu-1}. \tag{49}$$

In order, now, to determine the coefficients in (49) by the calculus of symbols, we note that

$$\frac{d\delta}{dE} = E^{-1}\Box$$

and, as is seen on differentiating $\delta_h = E^{\frac{h}{2}} - E^{-\frac{h}{2}}$ with respect to δ_h,

$$\frac{dE}{d\delta_h} = \frac{1}{h}E\,\Box_h^{-1},$$

so that

$$\frac{d\delta}{d\delta_h} = \frac{d\delta}{dE}\frac{dE}{d\delta_h} = \frac{1}{h}\Box\,\Box_h^{-1}.$$

But from this follows

$$\Box = h\Box_h\frac{d\delta}{d\delta_h}$$

and consequently, by (45),

$$\square = \square_h \sum_{\nu=0}^{\infty} \frac{1}{(2\nu)!}\left(\frac{1}{2h}\right)^{[2\nu+1]-1} \delta_h^{2\nu}. \tag{50}$$

If we insert this expression in (49), divide by $\square_h$, and multiply both sides by δ, taking account of (45), we find

$$\sum_{\nu=0}^{\infty} \frac{1}{(2\nu)!}\left(\frac{1}{2h}\right)^{[2\nu+1]-1} \delta_h^{2\nu}$$

$$= \left[\frac{1}{h}\delta_h + \frac{2}{3!}\left(\frac{1}{2h}\right)^{[3]}\delta^3_h + \frac{2}{5!}\left(\frac{1}{2h}\right)^{[5]}\delta^5_h + \ldots\right]\left[h\delta_h^{-1} + Q_2\delta_h + Q_4\delta_h^3 + \ldots\right],$$

whence the recurrence formula

$$\left.\begin{aligned} -Q_{2\nu} = {} & \frac{Q_{2\nu-2}}{3!}\left(\frac{1}{2h}\right)^{[3]-1} + \frac{Q_{2\nu-4}}{5!}\left(\frac{1}{2h}\right)^{[5]-1} + \ldots \\ & + \frac{Q_2}{(2\nu-1)!}\left(\frac{1}{2h}\right)^{[2\nu-1]-1} - \frac{2\nu h}{(2\nu+1)!}\left(\frac{1}{2h}\right)^{[2\nu+1]-1}, \end{aligned}\right\} \tag{51}$$

leading to the same values as found in No. 159. It is seen that $Q_{2\nu}$ like $P_{2\nu}$ has the form of $\dfrac{h^2-1}{h^{2\nu-1}}$ multiplied by a polynomial in h^2.

213. The first and second formula of Woolhouse's type are obtained by expanding

$$\square\delta^{-1} - h\square_h\,\delta_h^{-1}$$

and

$$\delta^{-1} - h\delta_h^{-1}$$

respectively in powers of D. We may do it simultaneously by expanding

$$\square_{2a}\,\delta^{-1} - h\square_{2ah}\,\delta_h^{-1}$$

and putting thereafter $a = \dfrac{1}{2}$, $a = 0$, respectively.

Now we have

$$\square_{2a}\delta^{-1} - h\square_{2ah}\delta_h^{-1} = \frac{1}{2}\,\frac{e^{(\frac{1}{2}+a)D} + e^{(\frac{1}{2}-a)D}}{e^D - 1} - \frac{h}{2}\,\frac{e^{(\frac{1}{2}+a)hD} + e^{(\frac{1}{2}-a)hD}}{e^{hD}-1},$$

where the right-hand side may be developed by §13 (32), so that

$$\square_{2a}\delta^{-1}-h\square_{2ah}\delta_h^{-1} = -\frac{1}{2}\sum_{\nu=1}^{\infty}\frac{h^\nu-1}{\nu!}[B_\nu(\tfrac{1}{2}+a)+B_\nu(\tfrac{1}{2}-a)]\,D^{\nu-1}. \quad (52)$$

If, in this, we put $a = \frac{1}{2}$, we get

$$\square\,\delta^{-1} = h\square_h\,\delta_h^{-1} - \sum_{\nu=2}^{\infty}\frac{h^\nu-1}{\nu!}B_\nu D^{\nu-1}, \quad (53)$$

the factor left out being $(E^{hm}-1)$; for $h = n$ the formula is seen to be identical with §15 (17).

If, on the other hand, we put $a = 0$, we find

$$\delta^{-1} = h\delta_h^{-1} - \sum_{\nu=2}^{\infty}\frac{h^\nu-1}{\nu!}B_\nu(\tfrac{1}{2})D^{\nu-1}, \quad (54)$$

the factor left out being $(E^{hm-\frac{1}{2}} - E^{-\frac{1}{2}})$. This formula is, for $h = n$, identical with §15 (21).

214. The operations we have dealt with may, of course, be repeated, and if we form the mth power of the left- and right-hand sides of the preceeding formulas, we obtain the corresponding developments.

The general expansions of D^m in differences, and of Δ^m in differential coefficients, are however better obtained as follows:

We have

$$a^{(\nu)} = \sum_{m=0}^{\infty}\frac{a^m}{m!}D^m\,0^{(\nu)},$$

where, in order to avoid a variable upper limit of summation, we have summed to infinity, this being permissible, as $D^m0^{(\nu)}$ vanishes for $m>\nu$.

If we insert this expression in (11), we get

$$E^a = \sum_{\nu=0}^{\infty}\frac{\Delta^\nu}{\nu!}\sum_{m=0}^{\infty}\frac{a^m}{m!}D^m0^{(\nu)} = \sum_{m=0}^{\infty}\frac{a^m}{m!}\sum_{\nu=0}^{\infty}\frac{D^m0^{(\nu)}}{\nu\;!}\Delta^\nu,$$

and hence, on comparing the coefficients of the same powers of a in this expression and (13),

$$D^m = \sum_{\nu=m}^{\infty} \frac{D^m 0^{(\nu)}}{\nu!} \Delta^\nu \tag{55}$$

or §7(17).

If, in the same way, we put

$$a^{[\nu]} = \sum_{m=0}^{\infty} \frac{a^m}{m!} D^m 0^{[\nu]},$$

we obtain from (19)

$$E^a = \sum_{\nu=0}^{\infty} \frac{\delta^\nu}{\nu!} \sum_{m=0}^{\infty} \frac{a^m}{m!} D^m 0^{[\nu]} = \sum_{m=0}^{\infty} \frac{a^m}{m!} \sum_{\nu=0}^{\infty} \frac{D^m 0^{[\nu]}}{\nu!} \delta^\nu$$

and on comparing with (13)

$$D^m = \sum_{\nu=m}^{\infty} \frac{D^m 0^{[\nu]}}{\nu!} \delta^\nu, \tag{56}$$

containing both §7 (23) and §7 (30).

We now put, in (30),

$$a^{[2\nu+1]-1} = \sum_{s=0}^{\infty} \frac{a^{2s}}{(2s)!} D^{2s} 0^{[2\nu+1]-1}$$

and thus obtain

$$\square_{2a} = \sum_{\nu=0}^{\infty} \frac{\square \delta^{2\nu}}{(2\nu)!} \sum_{s=0}^{\infty} \frac{a^{2s}}{(2s)!} D^{2s} 0^{[2\nu+1]-1} = \sum_{s=0}^{\infty} \frac{a^{2s}}{(2s)!} \sum_{\nu=0}^{\infty} \frac{D^{2s} 0^{[2\nu+1]-1}}{(2\nu)!} \square \delta^{2\nu}$$

and, on comparison with

$$\square_{2a} = \sum_{s=0}^{\infty} \frac{a^{2s}}{(2s)!} D^{2s},$$

$$D^{2s} = \sum_{\nu=s}^{\infty} \frac{D^{2s} 0^{[2\nu+1]-1}}{(2\nu)!} \square \delta^{2\nu} = \sum_{\nu=s}^{\infty} \frac{D^{2s+1} 0^{[2\nu+1]}}{(2\nu)!\,(2s+1)} \square \delta^{2\nu} \tag{57}$$

or §7(27).

If, next, we put in (26)

$$a^{[2\nu+2]-1} = \sum_{s=1}^{\infty} \frac{a^{2s-1}}{(2s-1)!} D^{2s-1} 0^{[2\nu+2]-1},$$

we find

$$\frac{1}{2}\delta_{2a} = \sum_{\nu=0}^{\infty} \frac{\Box\delta^{2\nu+1}}{(2\nu+1)!} \sum_{s=1}^{\infty} \frac{a^{2s-1}}{(2s-1)!} D^{2s-1}0^{[2\nu+2]-1}$$

$$= \sum_{s=1}^{\infty} \frac{a^{2s-1}}{(2s-1)!} \sum_{\nu=0}^{\infty} \frac{D^{2s-1}0^{[2\nu+2]-1}}{(2\nu+1)!} \Box\delta^{2\nu+1}$$

and, on comparing with

$$\frac{1}{2}\delta_{2a} = \sum_{s=1}^{\infty} \frac{a^{2s-1}}{(2s-1)!} D^{2s-1},$$

$$D^{2s-1} = \sum_{\nu=s-1}^{\infty} \frac{D^{2s-1}0^{[2\nu+2]-1}}{(2\nu+1)!} \Box\delta^{2\nu+1}$$

or

$$D^{2s-1} = \sum_{\nu=s}^{\infty} \frac{D^{2s-1}0^{[2\nu]-1}}{(2\nu-1)!} \Box\delta^{2\nu-1} = \sum_{\nu=s}^{\infty} \frac{D^{2s}0^{[2\nu]}}{(2\nu-1)!\,2s} \Box\delta^{2\nu-1}, \qquad (58)$$

being identical with §7 (26).

215. The differences, expressed in terms of differential coefficients, are obtained in a similar way. We content ourselves with a single example.

If, in (13), we put

$$a^{\nu} = \sum_{m=0}^{\infty} \frac{a^{(m)}}{m!} \Delta^m 0^{\nu},$$

we obtain

$$E^a = \sum_{\nu=0}^{\infty} \frac{D^{\nu}}{\nu!} \sum_{m=0}^{\infty} \frac{a^{(m)}}{m!} \Delta^m 0^{\nu} = \sum_{m=0}^{\infty} \frac{a^{(m)}}{m!} \sum_{\nu=0}^{\infty} \frac{\Delta^m 0^{\nu}}{\nu!} D^{\nu}$$

and, on comparison with (11),

$$\Delta^m = \sum_{\nu=m}^{\infty} \frac{\Delta^m 0^{\nu}}{\nu!} D^{\nu} \qquad (59)$$

or §7(46).

216. If a summation-symbol as, for instance, Δ^{-1}, is repeated, the indetermination is increased, as Δ^{-m} need only be of such a nature that $\Delta^m \Delta^{-m} = 1$ (compare the remarks in No. 195).

This indetermination vanishes, however, if we agree that the summation shall, at each repeated summation, begin with the same argument α. With this convention, Δ^{-1} has the same meaning as $\Sigma \equiv \sum_{\alpha}^{x-1}$ in §11.

As an example we will derive the Gaussian formulas for repeated integration. We obtain from (56)

$$D^2 = \sum_{s=1}^{\infty} \frac{D^2 0^{[2s]}}{(2s)!} \delta^{2s}.$$

If, now, we put

$$D^{-2} = -\sum_{\nu=0}^{\infty} N_{2\nu} \delta^{2\nu-2},$$

we obtain, on multiplying together the two equations

$$-1 = \left(\frac{D^2 0^{[2]}}{2!}\delta^2 + \frac{D^2 0^{[4]}}{4!}\delta^4 + \ldots\right)(N_0 \delta^{-2} + N_2 + N_4 \delta^2 + \ldots)$$

and therefore, in addition to $N_0 = -1$, the recurrence formula

$$-N_{2\nu} = N_{2\nu-2}\frac{D^2 0^{[4]}}{4!} + N_{2\nu-4}\frac{D^2 0^{[6]}}{6!} + \ldots + N_2 \frac{D^2 0^{[2\nu]}}{(2\nu)!} - \frac{D^2 0^{[2\nu+2]}}{(2\nu+2)!}, \quad (60)$$

leading to the same values of the $N_{2\nu}$ as found in No. 130 by another method.

We have, therefore,

$$\left.\begin{aligned} D^{-2} &= \delta^{-2} - \sum_{\nu=1}^{\infty} N_{2\nu} \delta^{2\nu-2} \\ &= E\Delta^{-2} - \sum_{\nu=1}^{\infty} N_{2\nu} \delta^{2\nu-2}, \end{aligned}\right\} \quad (61)$$

the indetermination consisting in introducing a linear function $ax + b$, if this operation is applied to $f(x)$.

The constants a and b are determined, if we require that integration and summation shall commence with the argument 0.

The constant a is introduced by the first integration and must, therefore, be determined by (36), as (61) is also obtained by applying (36) twice, eliminating $\square^2$ by $\square^2 = 1 + \frac{\delta^2}{4}$. Now (36) may be written

$$\int f(x)dx = \Sigma f(x) + \frac{1}{2}f(x) + \sum_{\nu=1}^{\infty} M_{2\nu}\square\,\delta^{2\nu-1}f(x) + a.$$

Assuming 0 as the lower limit of integration and summation, and putting $x = 0$, we have

$$a = -\frac{1}{2}f(0) - \sum_{\nu=1}^{\infty} M_{2\nu}\square\,\delta^{2\nu-1}f(0).$$

In order to determine b, we now write (61) in the form

$$\int\!\!\int f(x)dx^2 = \Sigma^2 f(x+1) - \sum_{\nu=1}^{\infty} N_{2\nu}\delta^{2\nu-2}f(x) + ax + b.$$

If, in this, we put $x = 0$ and remember that the lower limits of integration and summation are zero, we find

$$b = \sum_{\nu=1}^{\infty} N_{2\nu}\delta^{2\nu-2}f(0).$$

With the values found for a and b we have finally

$$\left.\begin{aligned}\int_0^x\!\!\int_0^x f(x)dx^2 = \Sigma^2 f(x+1) - x\left[\frac{1}{2}f(0) + \sum_{\nu=1}^{\infty} M_{2\nu}\square\,\delta^{2\nu-1}f(0)\right] \\ - \sum_{\nu=1}^{\infty} N_{2\nu}\Big[\delta^{2\nu-2}f(x)\Big]_0^x,\end{aligned}\right\} \quad (62)$$

agreeing with §12 (44), if in the latter formula we let $r \to \infty$.

217. In order to derive the first Gaussian formula for repeated integration, we note that, according to (57),

$$D^2 = \sum_{\nu=1}^{\infty} \frac{D^3 0^{[2\nu+1]}}{(2\nu)!\,3}\square\,\delta^{2\nu}.$$

If, now, we put

$$D^{-2} = -\sum_{\nu=0}^{\infty} J_{2\nu}\square\,\delta^{2\nu-2}$$

and multiply the two equations by each other, we have

$$-\frac{1}{1+\frac{\delta^2}{4}} = \left(\frac{D^3 0^{[3]}}{2!\,3}\delta^2 + \frac{D^3 0^{[5]}}{4!\,3}\delta^4 + \ldots\right)(J_0\delta^{-2} + J_2 + J_4\delta^2 + \ldots)$$

and, on comparing the coefficients of $\delta^{2\nu}$ in the expansions of both sides of the equation, $J_0 = -1$ and the recurrence formula

$$\left.\begin{aligned} -J_{2\nu} &= J_{2\nu-2}\frac{D^3 0^{[5]}}{4!\,3} + J_{2\nu-4}\frac{D^3 0^{[7]}}{6!\,3} + \ldots \\ &+ J_2\frac{D^3 0^{[2\nu+1]}}{(2\nu)!\,3} - \frac{D^3 0^{[2\nu+3]}}{(2\nu+2)!\,3} + \frac{(-1)^\nu}{2^{2\nu}}, \end{aligned}\right\}\quad(63)$$

leading to the same values as found in No. 129 by another method.

We therefore have, to begin with,

$$D^{-2} = \Box\delta^{-2} - \sum_{\nu=1}^{\infty} J_{2\nu}\Box\delta^{2\nu-2}. \quad (64)$$

If we apply this formula to $f(x - \frac{1}{2})$, and note that

$$\Box\delta^{-2}f(x - \tfrac{1}{2}) = \left(1 + \frac{\Delta}{2}\right)\Delta^{-2}f(x) = \Sigma^2 f(x) + \tfrac{1}{2}\Sigma f(x),$$

we have

$$\iint f(x-\tfrac{1}{2})dx^2 = \Sigma^2 f(x) + \tfrac{1}{2}\Sigma f(x) - \sum_{\nu=1}^{\infty} J_{2\nu}\Box\delta^{2\nu-2}f(x-\tfrac{1}{2}) + ax + b. \quad (65)$$

Assuming, as before, that the lower limits of integration and summation are zero, we find for $x = 0$

$$b = \sum_{\nu=1}^{\infty} J_{2\nu}\Box\delta^{2\nu-2}f(-\tfrac{1}{2}),$$

while a is determined in a similar way by (34) which we write in the form

$$\int f(x - \tfrac{1}{2})dx = \Sigma f(x) + \sum_{\nu=1}^{\infty} K_{2\nu}\delta^{2\nu-1}f(x - \tfrac{1}{2}) + a,$$

whence

$$a = -\sum_{\nu=1}^{\infty} K_{2\nu}\delta^{2\nu-1}f(-\tfrac{1}{2}).$$

If, finally, we insert in (65) the values found for a and b, we obtain

$$\left.\begin{aligned}\int_0^x\int_0^x f(x-\tfrac{1}{2})dx^2 &= \int_{-\frac{1}{2}}^{x-\frac{1}{2}}\int_{-\frac{1}{2}}^{x} f(x)dx^2 \\ = \Sigma^2 f(x) + \tfrac{1}{2}\Sigma f(x) - x\sum_{\nu=1}^{\infty} K_{2\nu}\delta^{2\nu-1}f(-\tfrac{1}{2}) &- \sum_{\nu=1}^{\infty} J_{2\nu}\Big[\Box\delta^{2\nu-2}f(x)\Big]_{-\frac{1}{2}}^{x-\frac{1}{2}},\end{aligned}\right\}\quad(66)$$

agreeing with §12 (36), if in the latter formula we let $r \to \infty$.

218. The special formula of Euler's type for repeated summation or integration, which we derived in No. 154, may be obtained by the calculus of symbols on expanding $\nabla^{-2}E^{-1}$ in powers of D.

On account of the relation

$$\begin{aligned}\frac{e^{-x}}{(1-e^{-x})^2} &= -D\frac{1}{e^x-1} \\ &= -\sum_0^{\infty}\frac{B_\nu}{\nu!}(\nu-1)x^{\nu-2} \\ &= x^{-2} - \sum_2^{\infty}\frac{B_\nu}{\nu}\frac{x^{\nu-2}}{(\nu-2)!}\end{aligned}$$

we have

$$\begin{aligned}\nabla^{-2}E^{-1} &= \frac{E^{-1}}{(1-E^{-1})^2} = \frac{e^{-D}}{(1-e^{-D})^2} \\ &= D^{-2} - \sum_{\nu=2}^{\infty}\frac{B_\nu}{\nu}\frac{D^{\nu-2}}{(\nu-2)!}\end{aligned}$$

or

$$D^{-2} = \nabla^{-2}E^{-1} + \sum_{\nu=2}^{\infty}\frac{B_\nu}{\nu}\frac{D^{\nu-2}}{(\nu-2)!}, \qquad (67)$$

agreeing with §14 (22) for $r \to \infty$; for we see, on letting $x \to \infty$, that both constants in the linear function, introduced by the integration, vanish.

§19. Interpolation with Several Variables[1]

219. Confining ourselves, to begin with, to functions of two variables, we denote a function of x and y by $f(x; y)$; not, as is customary in analysis, by $f(x, y)$ which, according to the notation of No. 16, means $\dfrac{f(x) - f(y)}{x - y}$.

We assume further, for the sake of simplicity, that the data are given in the following form, which we shall call *the function-table:*

x	$y = b_0$	b_1	b_2	...
a_0	$f(a_0; b_0)$	$f(a_0; b_1)$	$f(a_0; b_2)$	...
a_1	$f(a_1; b_0)$	$f(a_1; b_1)$	$f(a_1; b_2)$	...
a_2	$f(a_2; b_0)$	$f(a_2; b_1)$	$f(a_2; b_2)$	...
.	.	.	.	
	.	.	.	
.	.	.	.	

A perfectly safe way of interpolating for $f(x; y)$ in this table is evidently the following one. For the particular value of x under consideration we interpolate, in succession, for the values

$$f(x; b_0),\ f(x; b_1),\ f(x; b_2),\ \ldots . \tag{1}$$

At each of these interpolations we use one particular column of the table; these interpolations are, therefore, interpolations with one variable, and the accuracy may be controlled by the remainder-term in the usual way.

But the sequence (1) is evidently a table with one single argument of $f(x; y)$, considered as a function of y and with a constant x. In this table we may, therefore, interpolate for $f(x; y)$ in the usual way, controlling the accuracy by means of the remainder-term.

It might, therefore, seem that there is no need for developing a

[1]See further the following two papers which only came to my notice after the book had been sent to press:

S. Narumi: Some formulas in the theory of interpolation of many independent variables; Tôhoku Mathematical Journal, vol. 18 (1920), p. 309.

L. Neder: Interpolationsformeln für Funktionen mehrerer Argumente; Skandinavisk Aktuarietidskrift, 1926, p. 59.

separate theory of interpolation with several variables. There are, however, several reasons why this should be done. In many cases we are not satisfied with obtaining $f(x; y)$ for particular values of x and y, but want an approximate expression for $f(x; y)$ as a function of x and y. The extent of the given function-table must also be taken into consideration, as it is not always possible to interpolate for the values (1) with uniform accuracy. In any case we shall assume that we possess the values of $f(x; y)$ required in order to obtain a form useful practically for the remainder-term. From this follows that the points $(a_\nu; b_\mu)$ used for the interpolation cannot be chosen quite arbitrarily in the plane but—as has already been assumed in the function-table above—belong to the points of section of two systems of parallel lines of which one is parallel to the x-axis, the other to the y-axis, while the distances between the parallels are quite arbitrary.

220. If a table of $f(x; y)$ is given, divided differences may be formed with respect to x and y, separately or simultaneously. The divided difference with respect to x, formed with the arguments $a_0, a_1, \ldots a_\nu$, and with respect to y, formed with the arguments $b_0, b_1, \ldots b_\mu$, is of the order $\nu + \mu$, and is denoted by

$$f_{\nu\mu} \equiv f(a_0 \ldots a_\nu; b_0 \ldots b_\mu). \tag{2}$$

It is indifferent, in which order the successive differences are formed; for if the symbol θ_p (see No. 18) acts on x_0 alone, and the symbol θ'_q on y_0 alone, it is seen by a simple calculation that

$$\theta_p\theta'_q f(x_0; y_0) = \theta'_q\theta_p f(x_0; y_0). \tag{3}$$

In dealing with more variables than one, it is of great importance to abbreviate the notation as much as possible, and we shall therefore write

$$\left.\begin{array}{c} x_\nu = (x - a_0) \ldots (x - a_{\nu-1}); y_\nu = (y - b_0) \ldots (y - b_{\nu-1}); \\ x_0 = y_0 = 1. \end{array}\right\} \tag{4}$$

We have then, by Newton's formula with divided differences

$$f(x; y) = \sum_{\nu=0}^{n} x_\nu f(a_0 \ldots a_\nu; y) + x_{n+1} f(x, a_0 \ldots a_n; y).$$

But by the same formula

$$f(a_0 \ldots a_\nu; y) = \sum_{\mu=0}^{m} y_\mu f_{\nu\mu} + y_{m+1} f(a_0 \ldots a_\nu; y, b_0 \ldots b_m),$$

so that

$$f(x; y) = \sum_{\nu=0}^{n} \sum_{\mu=0}^{m} x_\nu y_\mu f_{\nu\mu} + R, \tag{5}$$

$$R = y_{m+1} \sum_{\nu=0}^{n} x_\nu f(a_0 \ldots a_\nu; y, b_0 \ldots b_m) + x_{n+1} f(x, a_0 \ldots a_n; y). \tag{6}$$

Now it is easy to simplify the expression for R; for we have by Newton's formula with divided differences

$$f(x; y, b_0 \ldots b_m)$$

$$= \sum_{\nu=0}^{n} x_\nu f(a_0 \ldots a_\nu; y, b_0 \ldots b_m) + x_{n+1} f(x, a_0 \ldots a_n; y, b_0 \ldots b_m),$$

and hence, on eliminating $\sum_{\nu=0}^{n}$ between this equation and (6),

$$\left.\begin{aligned} R &= x_{n+1} f(x, a_0 \ldots a_n; y) + y_{m+1} f(x; y, b_0 \ldots b_m) \\ &\quad - x_{n+1} y_{m+1} f(x, a_0 \ldots a_n; y, b_0 \ldots b_m). \end{aligned}\right\} \tag{7}$$

A practical form of the remainder-term may be obtained by noting that

$$f(x, a_0 \ldots a_n; y, b_0 \ldots b_m) = \frac{1}{(n+1)!\,(m+1)!} D_\xi^{n+1} D_\eta^{m+1} f(\xi; \eta), \tag{8}$$

D_ξ and D_η denoting partial differentiation. For we have first

$$f(x, a_0 \ldots a_n; y, b_0 \ldots b_m) = \frac{1}{(m+1)!} D_\eta^{m+1} f(x, a_0 \ldots a_n; \eta),$$

η depending on x; but as D_ξ denotes a *partial* differentiation, we may write

$$f(x, a_0 \ldots a_n; \eta) = \frac{1}{(n+1)!} D_\xi^{n+1} f(\xi; \eta).$$

We therefore obtain from (7)

$$R = \frac{x_{n+1}}{(n+1)!} D_\xi^{n+1} + \frac{y_{m+1}}{(m+1)!} D_\eta^{m+1} - \frac{x_{n+1}}{(n+1)!} \frac{y_{m+1}}{(m+1)!} D_\xi^{n+1} D_\eta^{m+1}, \quad (9)$$

where we have, for abbreviation, left out $f(\xi; y)$, $f(x; \eta)$ and $f(\xi; \eta)$. As the actual values of ξ and η are unknown, it can hardly cause any misunderstanding that these numbers are not assumed to have the same values in the third term of R as in the two first ones.

The expression (9) is, of course, not so simple as in the case of interpolation with one variable; but it is, in reality, simpler than what might have been anticipated in interpolating with two variables, and perfectly suited for numerical calculations. It should be noted that the third term of R (apart from the sign) is not the product, but the *symbolical* product of the two first terms.

As the most important cases, practically, of (5) and (9) we note, for $n = m = 1$

$$\left.\begin{aligned} f(x; y) &= f_{00} + x_1 f_{10} + y_1 f_{01} + x_1 y_1 f_{11} + R, \\ R &= \frac{x_2}{2} D_\xi^2 + \frac{y_2}{2} D_\eta^2 - \frac{x_2}{2}\frac{y_2}{2} D_\xi^2 D_\eta^2 \end{aligned}\right\} (10)$$

and for $n = m = 2$

$$\left.\begin{aligned} f(x; y) &= f_{00} + x_1 f_{10} + x_2 f_{20} \\ &\quad + y_1 (f_{01} + x_1 f_{11} + x_2 f_{21}) \\ &\quad + y_2 (f_{02} + x_1 f_{12} + x_2 f_{22}) + R, \\ R &= \frac{x_3}{6} D_\xi^3 + \frac{y_3}{6} D_\eta^3 - \frac{x_3}{6}\frac{y_3}{6} D_\xi^3 D_\eta^3 . \end{aligned}\right\} (11)$$

221. Amongst the general properties of the polynomial in (5) we note that it is of degree n in x and of degree m in y. Further, that for $x = a_\nu$ $(0 \leq \nu \leq n)$, $y = b_\mu$ $(0 \leq \mu \leq m)$ it assumes the value $f(a_\nu; b_\mu)$, as the remainder-term vanishes under these circumstances. The points used in the interpolation are *all* the points $(a_\nu; b_\mu)$ for which $0 \leq \nu \leq n$, $0 \leq \mu \leq m$. These points are the intersection points of two systems of parallels, of which $n + 1$ are parallel to the y-axis and $m + 1$ to the x-axis.

It is clear that interpolation by (5) is equivalent to the method described above of interpolating first for one variable and then for the other, provided that, in interpolating for x, we use the nth difference in each y-column, and that, in interpolating for y, the mth difference is used throughout. It is also clear that we obtain the same result, whether we interpolate first for x and then for y, or *vice versa*.

If we only want the value of $f(x; y)$ for one particular value of x and y, it is easier to interpolate first for one and then for the other variable, than to employ (5), see No. 236 and No. 240; but even then, our analysis has shown that we may content ourselves with applying the remainder-term (9), instead of examining the remainder-term at each individual interpolation.

222. If in (5) we put $a_\nu = 0, b_\mu = 0$ and write, for abbreviation, $D_u^\nu D_v^\mu$ instead of $D_{u=0}^\nu D_{v=0}^\mu f(u; v)$, we get

$$\left.\begin{aligned} f(x;y) &= \sum_{\nu=0}^{n} \sum_{\mu=0}^{m} \frac{x^\nu}{\nu!}\frac{y^\mu}{\mu!} D_u^\nu D_v^\mu + R, \\ R &= \frac{x^{n+1}}{(n+1)!} D_\xi^{n+1} + \frac{y^{m+1}}{(m+1)!} D_\eta^{m+1} - \frac{x^{n+1}}{(n+1)!}\frac{y^{m+1}}{(m+1)!} D_\xi^{n+1} D_\eta^{m+1} \end{aligned}\right\} \quad (12)$$

This formula is not identical with the usual form of Taylor's formula for two variables, the latter proceeding by *homogeneous* polynomials in x and y and having a different remainder-term (No. 232).

Next let us, in (5), put $a_\nu = \nu$, $b_\mu = \mu$. Using the notation $\overset{\nu}{\Delta}_u \overset{\mu}{\Delta}_v$ in a sense analogous with $D_u^\nu D_v^\mu$, and not to be confounded with the differences for arbitrary intervals occurring in earlier sections, we find

$$\left.\begin{aligned} f(x;y) &= \sum_{\nu=0}^{n} \sum_{\mu=0}^{m} \frac{x^{(\nu)}}{\nu!}\frac{y^{(\mu)}}{\mu!} \overset{\nu}{\Delta}_u \overset{\mu}{\Delta}_v + R, \\ R &= \frac{x^{(n+1)}}{(n+1)!} D_\xi^{n+1} + \frac{y^{(m+1)}}{(m+1)!} D_\eta^{m+1} - \frac{x^{(n+1)}}{(n+1)!}\frac{y^{(m+1)}}{(m+1)!} D_\xi^{n+1} D_\eta^{m+1} \end{aligned}\right\} \quad (13)$$

being the interpolation-formula with descending differences for two variables. We need hardly write down the corresponding formula with ascending differences.

223. Stirling's and Bessel's formulas with two variables may, of course, be obtained directly from (5), but it is easier to derive them from the formulas with one variable, by means of the following considerations.

If we follow the line of argument leading to (5) with the remainder-term (7), it appears, that we have made no use at all of the circumstances that x_ν is the special function $(x - a_0)\,(x - a_1) \ldots (x - a_{\nu-1})$ and $f(a_0, a_1, \ldots a_\nu)$ a divided difference. In reality we have only made use of the fact that $f(x)$ has the *form*

$$f(x) = \sum_{\nu=0}^{n} x_\nu f(a_0 \ldots a_\nu) + x_{n+1} f(x, a_0 \ldots a_n), \tag{14}$$

the x_ν being certain given functions of $x, a_0, \ldots a_{\nu-1}$, and the $f(a_0 \ldots a_\nu)$ functions of $a_0, \ldots a_\nu$ which are obtained by performing certain operations on $f(x)$. Formula (5) is therefore valid on these more general assumptions.

We now denote the operation by which $f(a_0 \ldots a_\nu)$ is derived from $f(x)$, by

$$\theta_u^\nu f(u) = f(a_0 \ldots a_\nu). \tag{15}$$

The operation θ^ν need not, therefore, necessarily mean the ν times repeated operation θ, but includes, for instance, such operations as $\theta^\nu = \square\,\delta^\nu$, etc. If the operation θ^ν is *distributive* and *commutative*, (5) may be written in the form of the symbolical product

$$f(x;y) = \sum_{\nu=0}^{n} x_\nu\,\theta_u^\nu \cdot \sum_{\mu=0}^{m} y_\mu\,\theta_v^\mu + R. \tag{16}$$

This is perhaps most easily seen by noting that, if by a_{n+1} we understand x, (14) may be written

$$f(x) = \sum_{\nu=0}^{n+1} x_\nu\,\theta_u^\nu\,f(u)$$

whence

$$f(x;y) = \sum_{\nu=0}^{n+1} x_\nu\,\theta_u^\nu \cdot \sum_{\mu=0}^{m+1} y_\mu\,\theta_v^\mu$$

$$= \sum_{\nu=0}^{n} x_\nu\,\theta_u^\nu \cdot \sum_{\mu=0}^{m} y_\mu\,\theta_v^\mu + R$$

where

$$\left.\begin{aligned} R = x_{n+1}\,\theta_u^{n+1} \sum_{\mu=0}^{m} y_\mu\,\theta_v^{\mu} + y_{m+1}\,\theta_v^{m+1} \sum_{\nu=0}^{n} x_\nu\,\theta_u^{\nu} \\ + x_{n+1}\,y_{m+1}\,\theta_u^{n+1}\,\theta_v^{m+1}. \end{aligned}\right\}$$

Removing now the sums by

$$\sum_{\nu=0}^{n} x_\nu\,\theta_u^{\nu} f(u;v) = f(x;v) - x_{n+1}\,\theta_u^{n+1} f(u;v),$$

$$\sum_{\mu=0}^{m} y_\mu\,\theta_v^{\mu} f(u;v) = f(u;y) - y_{m+1}\,\theta_v^{m+1} f(u;v),$$

we have the short form for the remainder-term

$$R = x_{n+1}\,\theta_u^{n+1} + y_{m+1}\,\theta_v^{m+1} - x_{n+1}\,y_{m+1}\,\theta_u^{n+1}\,\theta_v^{m+1}.$$

224. It is clear that in this way we obtain three types of interpolation-formulas with central differences, as each of the symbolical factors in (16) may be either of Stirling's or of Bessel's type. Thus, the formula of the type Stirling-Stirling is the symbolical product of the two formulas

$$\left.\begin{aligned} &\sum_{\nu=0}^{n} \left[\frac{x^{[2\nu]-1}}{(2\nu-1)!}\,\Box\,\delta_u^{2\nu-1} + \frac{x^{[2\nu]}}{(2\nu)!}\,\delta_u^{2\nu}\right] \\ \text{and} \\ &\sum_{\mu=0}^{m} \left[\frac{y^{[2\mu]-1}}{(2\mu-1)!}\,\Box\,\delta_v^{2\mu-1} + \frac{y^{[2\mu]}}{(2\mu)!}\,\delta_v^{2\mu}\right] \end{aligned}\right\} \quad (17)$$

with the remainder-term

$$R = \frac{x^{[2n+2]-1}}{(2n+1)!}D_\xi^{2n+1} + \frac{y^{[2m+2]-1}}{(2m+1)!}D_\eta^{2m+1} - \frac{x^{[2n+2]-1}}{(2n+1)!}\,\frac{y^{[2m+2]-1}}{(2m+1)!}D_\xi^{2n+1}\,D_\eta^{2m+1}. \quad (18)$$

We have, for instance, for $n = m = 1$

$$\left.\begin{aligned} f(x;y) &= 1 + x\,\Box\,\delta_u + \frac{x^2}{2}\,\delta_u^2 \\ &+ y\left(\Box\,\delta_v + x\,\Box\,\delta_u\,\Box\,\delta_v + \frac{x^2}{2}\,\delta_u^2\,\Box\,\delta_v\right) \\ &+ \frac{y^2}{2}\left(\delta_v^2 + x\,\Box\,\delta_u\,\delta_v^2 + \frac{x^2}{2}\,\delta_u^2\,\delta_v^2\right) + R, \end{aligned}\right\} \quad (19)$$

$$R = \frac{x(x^2-1)}{6} D_\xi^3 + \frac{y(y^2-1)}{6} D_\eta^3 - \frac{x(x^2-1)}{6} \frac{y(y^2-1)}{6} D_\xi^3 D_\eta^3. \quad (20)$$

The factor left out in (19) is $f(u; v)$, so that the first term on the right, or 1, means $f(0; 0)$.

225. In the same way, the formula of the type Bessel-Bessel is the symbolical product of the two formulas

$$\sum_{\nu=0}^{n} \left[\frac{x^{[2\nu+1]-1}}{(2\nu)!} \square \, \delta_u^{2\nu} + \frac{x^{[2\nu+1]}}{(2\nu+1)!} \delta_u^{2\nu+1} \right]$$

and (21)

$$\sum_{\mu=0}^{m} \left[\frac{y^{[2\mu+1]-1}}{(2\mu)!} \square \, \delta_v^{\mu} + \frac{y^{[2\mu+1]}}{(2\mu+1)!} \delta_v^{2\mu+1} \right]$$

with the factor $f(u + \frac{1}{2}; v + \frac{1}{2})$ left out, and the remainder-term

$$R = \frac{x^{[2n+3]-1}}{(2n+2)!} D_\xi^{2n+2} + \frac{y^{[2m+3]-1}}{(2m+2)!} D_\eta^{2m+2} - \frac{x^{[2n+3]-1}}{(2n+2)!} \frac{y^{[2m+3]-1}}{(2m+2)!} D_\xi^{2n+2} D_\eta^{2m+2}. \quad (22)$$

We have, for instance, for $n = m = 1$

$$\left.\begin{aligned} f(x + \tfrac{1}{2}; y + \tfrac{1}{2}) &= \square_u \square_v + x\delta_u \square_v + \frac{x^2 - \frac{1}{4}}{2} \square \, \delta_u^2 \square_v + \frac{x(x^2 - \frac{1}{4})}{6} \delta_u^3 \square_v \\ &+ y\left[\square_u \delta_v + x\delta_u \delta_v + \frac{x^2 - \frac{1}{4}}{2} \square \, \delta_u^2 \delta_v + \frac{x(x^2 - \frac{1}{4})}{6} \delta_u^3 \delta_v \right] \\ &+ \frac{y^2 - \frac{1}{4}}{2} \left[\square_u \square \, \delta_v^2 + x\delta_u \square \, \delta_v^2 + \frac{x^2 - \frac{1}{4}}{2} \square \, \delta_u^2 \square \, \delta_v^2 + \frac{x(x^2 - \frac{1}{4})}{6} \delta_u^3 \square \, \delta_v^2 \right] \\ &+ \frac{y(y^2 - \frac{1}{4})}{6} \left[\square_u \delta_v^3 + x\,\delta_u \delta_v^3 + \frac{x^2 - \frac{1}{4}}{2} \square \, \delta_u^2 \delta_v^3 + \frac{x(x^2 - \frac{1}{4})}{6} \delta_u^3 \delta_v^3 \right] + R \end{aligned}\right\} (23)$$

with the factor $f(u + \frac{1}{2}; v + \frac{1}{2})$ left out, and the remainder-term

$$\left.\begin{aligned} R &= \frac{(x^2 - \frac{1}{4})(x^2 - \frac{9}{4})}{24} D_\xi^4 \\ &+ \frac{(y^2 - \frac{1}{4})(y^2 - \frac{9}{4})}{24} D_\eta^4 - \frac{(x^2 - \frac{1}{4})(x^2 - \frac{9}{4})}{24} \frac{(y^2 - \frac{1}{4})(y^2 - \frac{9}{4})}{24} D_\xi^4 D_\eta^4. \end{aligned}\right\} (24)$$

226. Finally the formula of the type Stirling-Bessel is obtained as the symbolical product of

$$\left.\begin{aligned}&\sum_{\nu=0}^{n}\left[\frac{x^{[2\nu]-1}}{(2\nu-1)!}\,\Box\,\delta_u^{2\nu-1}+\frac{x^{[2\nu]}}{(2\nu)!}\,\delta_u^{2\nu}\right]\\ &\text{and}\\ &\sum_{\mu=0}^{m}\left[\frac{y^{[2\mu+1]-1}}{(2\mu)!}\,\Box\,\delta_v^{2\mu}+\frac{y^{[2\mu+1]}}{(2\mu+1)!}\,\delta_v^{2\mu+1}\right]\end{aligned}\right\}\quad(25)$$

with the factor $f(u; v + \frac{1}{2})$ left out, and the remainder-term

$$R=\frac{x^{[2n+2]-1}}{(2n+1)!}D_\xi^{2n+1}+\frac{y^{[2m+3]-1}}{(2m+2)!}D_\eta^{2m+2}-\frac{x^{[2n+2]-1}}{(2n+1)!}\,\frac{y^{[2m+3]-1}}{(2m+2)!}D_\xi^{2n+1}D_\eta^{2m+2}. \quad (26)$$

We have, for instance, for $n = m = 1$

$$\left.\begin{aligned}f(x;y+\tfrac{1}{2}) &= \Box_v + x\Box\,\delta_u\,\Box_v + \frac{x^2}{2}\,\delta_u^2\,\Box_v\\ &\quad + y\left(\delta_v + x\Box\,\delta_u\,\delta_v + \frac{x^2}{2}\,\delta_u^2\,\delta_v\right)\\ &+ \frac{y^2-\frac{1}{4}}{2}\left(\Box\,\delta_v^2 + x\Box\,\delta_u\,\Box\,\delta_v^2 + \frac{x^2}{2}\,\delta_u^2\,\Box\,\delta_v^2\right)\\ &+ \frac{y(y^2-\frac{1}{4})}{6}\left(\delta_v^3 + x\Box\,\delta_u\,\delta_v^3 + \frac{x^2}{2}\,\delta_u^2\,\delta_v^3\right) + R\end{aligned}\right\}\quad(27)$$

with the factor $f(u; v + \frac{1}{2})$ left out, and the remainder-term

$$\left.\begin{aligned}R &= \frac{x(x^2-1)}{6}D_\xi^3 + \frac{(y^2-\frac{1}{4})\,(y^2-\frac{9}{4})}{24}D_\eta^4\\ &\quad - \frac{x(x^2-1)}{6}\,\frac{(y^2-\frac{1}{4})\,(y^2-\frac{9}{4})}{24}D_\xi^3\,D_\eta^4.\end{aligned}\right\}\quad(28)$$

227. We note, in particular, the formula for *interpolation to the* *middle.* It is obtained from the type Bessel-Bessel by putting $x = y = 0$ and is, therefore, the symbolical product of

$$\left.\begin{aligned}&\sum_{\nu=0}^{n} (-1)^{\nu} \frac{[1.3 \ldots (2\nu-1)]^2}{2^{2\nu}(2\nu)!} \square\, \delta_u^{2\nu}\\ &\text{and}\\ &\sum_{\mu=0}^{m} (-1)^{\mu} \frac{[1.3 \ldots (2\mu-1)]^2}{2^{2\mu}(2\mu)!} \square\, \delta_v^{2\mu}\end{aligned}\right\} \quad (29)$$

with the factor $f(u + \frac{1}{2}; v + \frac{1}{2})$ left out, and the remainder-term

$$\left.\begin{aligned}R = {} & (-1)^{n+1} \frac{[1.3 \ldots (2n+1)]^2}{2^{2n+2}(2n+2)!} D_\xi^{2n+2} + (-1)^{m+1} \frac{[1.3 \ldots (2m+1)]^2}{2^{2m+2}(2m+2)!} D_\eta^{2m+2}\\ & -(-1)^{n+m} \frac{[1.3 \ldots (2n+1)]^2}{2^{2n+2}(2n+2)!} \frac{[1.3 \ldots (2m+1)]^2}{2^{2m+2}(2m+2)!} D_\xi^{2n+2} D_\eta^{2m+2}.\end{aligned}\right\} \quad (30)$$

This formula is, in reality, fairly simple if written with numerical coefficients. Thus we have, for $n = m = 2$,

$$\left.\begin{aligned}f\left(\frac{1}{2};\frac{1}{2}\right) = {} & \square_u \square_v - \frac{1}{8} \square\, \delta_u^2 \square_v + \frac{3}{128} \square\, \delta_u^4 \square_v\\ & - \frac{1}{8}\left(\square_u \square \delta_v^2 - \frac{1}{8} \square \delta_u^2 \square \delta_v^2 + \frac{3}{128} \square \delta_u^4 \square \delta_v^2\right)\\ & + \frac{3}{128}\left(\square_u \square \delta_v^4 - \frac{1}{8} \square \delta_u^2 \square \delta_v^4 + \frac{3}{128} \square \delta_u^4 \square \delta_v^4\right) + R\end{aligned}\right\} \quad (31)$$

with the factor $f\left(u + \dfrac{1}{2}; v + \dfrac{1}{2}\right)$ left out, and the remainder-term

$$R = -\frac{5}{1024}(D_\xi^6 + D_\eta^6) - \left(\frac{5}{1024}\right)^2 D_\xi^6 D_\eta^6. \quad (32)$$

228. Formulas with two variables of Everett's type may be obtained in exactly the same way. We leave the details to the reader.

229. Interpolation-formulas with three or more variables may be obtained by the same method. We confine ourselves to stating these formulas in the general case; that is the case of divided differences, from which all the more special formulas may easily be derived.

If there are three variables, we begin by expanding in terms of one of these, and thereafter use the results obtained for two variables, that is (5) and (7). The Σ-terms occurring in the remainder-term are eliminated as above by Newton's formula, written in the form

$$\sum_{\nu=0}^{n} x_\nu f(a_0 \dots a_\nu) = f(x) - x_{n+1} f(x, a_0 \dots a_n).$$

In this way we find

$$f(x;y;z) = \sum_{\nu=0}^{n} \sum_{\mu=0}^{m} \sum_{\kappa=0}^{k} x_\nu y_\mu z_\kappa f_{\nu\mu\kappa} + R \tag{33}$$

or in symbolical form

$$f(x;y;z) = \sum_{\nu=0}^{n} x_\nu \theta_u^\nu \cdot \sum_{\mu=0}^{m} y_\mu \theta_v^\mu \cdot \sum_{\kappa=0}^{k} z_\kappa \theta_w^\kappa + R, \tag{34}$$

the factor $f(u;v;w)$ having been left out.

For the remainder-term we have, to begin with, the inconvenient form

$$\left.\begin{aligned} R = {} & x_{n+1} f(x, a_0 \dots a_n; y; z) + y_{m+1} f(x; y, b_0 \dots b_m; z) \\ & + z_{k+1} f(x; y; z, c_0 \dots c_k) - x_{n+1} y_{m+1} f(x, a_0 \dots a_n; y, b_0 \dots b_m; z) \\ & - x_{n+1} z_{k+1} f(x, a_0 \dots a_n; y; z, c_0 \dots c_k) - y_{m+1} z_{k+1} f(x; y, b_0 \dots b_m; z, c_0 \dots c_k) \\ & + x_{n+1} y_{m+1} z_{k+1} f(x, a_0 \dots a_n; y, b_0 \dots b_m; z, c_0 \dots c_k). \end{aligned}\right\} \tag{35}$$

In order to simplify this, we again abbreviate our notation, writing

$$\left.\begin{aligned} \bar{x}_{n+1} &= x_{n+1} f(x, a_0 \dots a_n; y; z) \\ &= \frac{x_{n+1}}{(n+1)!} D_\xi^{n+1}, \\ \bar{x}_{n+1} \bar{y}_{m+1} &= x_{n+1} y_{m+1} f(x, a_0 \dots a_n; y, b_0 \dots b_m; z) \\ &= \frac{x_{n+1}}{(n+1)!} \frac{y_{m+1}}{(m+1)!} D_\xi^{n+1} D_\eta^{m+1}, \\ \bar{x}_{n+1} \bar{y}_{m+1} \bar{z}_{k+1} &= x_{n+1} y_{m+1} z_{k+1} f(x, a_0 \dots a_n; y, b_0 \dots b_m; z, c_0 \dots c_k) \\ &= \frac{x_{n+1}}{(n+1)!} \frac{y_{m+1}}{(m+1)!} \frac{z_{k+1}}{(k+1)!} D_\xi^{n+1} D_\eta^{m+1} D_\zeta^{k+1} \end{aligned}\right\} \tag{36}$$

etc. In this notation the remainder-term assumes the more transparent form

$$\left.\begin{aligned} R = (\bar{x}_{n+1} + \bar{y}_{m+1} + \bar{z}_{k+1}) - (\bar{x}_{n+1}\bar{y}_{m+1} + \bar{x}_{n+1}\bar{z}_{k+1} + \bar{y}_{m+1}\bar{z}_{k+1}) \\ + \bar{x}_{n+1}\bar{y}_{m+1}\bar{z}_{k+1}. \end{aligned}\right\} \quad (37)$$

230. It is finally proved by induction that analogous formulas hold for N variables. If Σ means summation with regard to all the *variables* $x, y, z, \ldots$, we find for the remainder-term in the general case

$$R = \Sigma\, \bar{x}_{n+1} - \Sigma\, \bar{x}_{n+1}\bar{y}_{m+1} + \Sigma\, \bar{x}_{n+1}\bar{y}_{m+1}\bar{z}_{k+1} - \ldots \quad (38)$$

The number of terms in this formula is

$$N + \binom{N}{2} + \binom{N}{3} + \ldots + 1 = 2^N - 1.$$

This number increases rapidly with N, but is, on the whole, small in comparison with the number of principal terms in the development of $f(x; y; z; \ldots)$, the latter number being

$$(n + 1)\,(m + 1)\,(k + 1) \ldots \qquad (N \text{ factors}).$$

231. So far we have only considered certain interpolation-formulas, possessing remainder-terms of a particularly simple form. We are now going to briefly mention certain formulas of a more elastic nature, but with more complicated remainder-terms. We confine ourselves to the case of two variables and to the formulas with divided differences. The reader who has followed our preceding explanations will have no difficulty in performing for himself the extension to more variables, and the specialization to descending and central differences, etc.

If, in deriving (5), we do not expand all the $f(a_0 \ldots a_\nu\,; y)$ to the same number of terms, that is, if we make m depend on ν, we evidently obtain a more general formula. We may express this by writing m_ν instead of m, in which case, instead of (5), we obtain

$$f(x; y) = \sum_{\nu=0}^{n} \sum_{\mu=0}^{m_\nu} x_\nu\, y_\mu\, f_{\nu\mu} + R, \quad (39)$$

$$R = x_{n+1} f(x, a_0 \ldots a_n; y) + \sum_{\nu=0}^{n} x_\nu y_{m_\nu+1} f(a_0 \ldots a_\nu; y, b_0 \ldots b_{m_\nu}) \quad (40)$$

or for practical purposes

$$R = \frac{x_{n+1}}{(n+1)!} D_\xi^{n+1} + \sum_{\nu=0}^{n} \frac{x_\nu}{\nu!} \frac{y_{m_\nu+1}}{(m_\nu+1)!} D_\xi^\nu D_\eta^{m_\nu+1}. \quad (41)$$

This remainder-term is not so simple as (9), but not so complicated that it is impossible to apply it in practice, especially as compensation is obtained by the fact that there are fewer terms to calculate in (39) than in (5). The choice of the numbers m_ν and n depends on an examination of the separate terms in (41) which is, of course, an inconvenience. In practice it is, therefore, often preferable to apply (5) with the remainder-term (9) in such a way that the terms which can at once be seen not to influence the required decimals, are not actually calculated. The same remark applies to the more special formulas, such as (23), etc.

232. A particular case of (39) deserves special attention.[1] If we put $m_\nu = n - \nu$, we obtain immediately

$$f(x; y) = \sum_{\nu=0}^{n} \sum_{\mu=0}^{n-\nu} x_\nu y_\mu f_{\nu\mu} + R, \quad (42)$$

$$R = \sum_{\nu=0}^{n+1} \frac{x_\nu}{\nu!} \frac{y_{n-\nu+1}}{(n-\nu+1)!} D_\xi^\nu D_\eta^{n-\nu+1}. \quad (43)$$

This may more conveniently be written in the following form

$$\left.\begin{aligned} f(x;y) = f_{00} + (x_1 f_{10} + y_1 f_{01}) + (x_2 f_{20} + x_1 y_1 f_{11} + y_2 f_{02}) + \ldots \\ + (x_n f_{n0} + x_{n-1} y_1 f_{n-1,1} + \ldots + y_n f_{0n}) + R, \end{aligned}\right\} (44)$$

$$R = \frac{x_{n+1}}{(n+1)!} D_\xi^{n+1} + \frac{x_n}{n!} \frac{y_1}{1!} D_\xi^n D_\eta + \ldots + \frac{y_{n+1}}{(n+1)!} D_\eta^{n+1}. \quad (45)$$

An examination of the remainder-term shows that it vanishes at the following points

$$\left.\begin{array}{l} (a_0; b_n) \\ (a_0; b_{n-1}),\ (a_1; b_{n-1}) \\ (a_0; b_{n-2}),\ (a_1; b_{n-2}),\ (a_2; b_{n-2}) \\ \cdots\cdots\cdots\cdots\cdots\cdots \\ (a_0; b_0),\ (a_1; b_0),\ \ldots\ldots\ldots\ldots\ (a_n; b_0). \end{array}\right\}$$

[1] O. Biermann: Vorlesungen über Mathematische Näherungsmethoden, p. 138–144.

At these points the polynomial in (44) therefore assumes the values $f(a_\nu\,;b_\mu)$, and the number of constants in the polynomial is the same as the number of these points, or $\frac{1}{2}(n+1)(n+2)$. A polynomial of this form is, therefore, completely determined, if we require that it shall assume these values, and (44) is a means of constructing such a polynomial.

Taylor's formula for two variables is evidently obtained as a limiting case of (44).

233. We may write (5) in a form corresponding to Lagrange's interpolation-formula. For if we put

$$\left.\begin{aligned} P(x) &= (x-a_0)(x-a_1)\ldots(x-a_n),\ P_\nu(x)=\frac{P(x)}{x-a_\nu}, \\ \overline{P}(y) &= (y-b_0)(y-b_1)\ldots(y-b_m),\ \overline{P}_\mu(y)=\frac{\overline{P}(y)}{y-b_\mu}, \end{aligned}\right\}\quad(46)$$

the polynomial

$$\sum_{\nu=0}^{n}\sum_{\mu=0}^{m}\frac{P_\nu(x)\,\overline{P}_\mu(y)}{P_\nu(a_\nu)\,\overline{P}_\mu(b_\mu)}f(a_\nu;b_\mu)$$

will evidently for $x=a_\nu\ (0\le\nu\le n),\ y=b_\mu\ (0\le\mu\le m)$ assume the values $f(a_\nu;\,b_\mu)$. As the polynomial is of degree n in x, and of degree m in y, it must be identical with the polynomial on the right of (5). We therefore have

$$f(x;y)=\sum_{\nu=0}^{n}\sum_{\mu=0}^{m}\frac{P_\nu(x)\,\overline{P}_\mu(y)}{P_\nu(a_\nu)\,\overline{P}_\mu(b_\mu)}f(a_\nu;b_\mu)+R, \qquad (47)$$

the remainder-term being the same as in (5) and therefore identical with (7) or (9).

We shall call this formula *Lagrange's formula for two variables.* It is easily extended to three and more variables. It is very useful for deriving certain formulas of mechanical cubature, see §20.

234. As regards the practical calculation of the coefficients in the interpolation-formula, we may begin by forming, by means of the function-table, the following table which is called *the difference-table in* x.

$f(a_0; b_0)$	$f(a_0; b_1)$	$f(a_0; b_2)$	$\cdots$
$f(a_0, a_1; b_0)$	$f(a_0, a_1; b_1)$	$f(a_0, a_1; b_2)$	$\cdots$
$f(a_0, a_1, a_2; b_0)$	$f(a_0, a_1, a_2; b_1)$	$f(a_0, a_1, a_2; b_2)$	$\cdots$
$\vdots$	$\vdots$	$\vdots$	

The numbers in the first, second, etc., column are obtained by forming the divided differences of the values in the first, second, etc., *column* of the function-table.

Having thus filled up the difference-table in x, we form the following *difference-table in x and y* by forming the divided differences of the values in the first, second, etc., *line* of the difference-table in x.

$f(a_0; b_0)$	$f(a_0; b_0, b_1)$	$f(a_0; b_0, b_1, b_2)$	$\cdots$
$f(a_0, a_1; b_0)$	$f(a_0, a_1; b_0, b_1)$	$f(a_0, a_1; b_0, b_1, b_2)$	$\cdots$
$f(a_0, a_1, a_2; b_0)$	$f(a_0, a_1, a_2; b_0, b_1)$	$f(a_0, a_1, a_2; b_0, b_1, b_2)$	$\cdots$
$\vdots$	$\vdots$	$\vdots$	

But this table is identical with

f_{00}	f_{01}	f_{02}	$\cdots$
f_{10}	f_{11}	f_{12}	$\cdots$
f_{20}	f_{21}	f_{22}	$\cdots$
$\vdots$	$\vdots$	$\vdots$	

and therefore contains exactly the divided differences of which use is made in the interpolation-formula. It should be noted that *columns* in the table correspond to *lines* in (11), and *vice versa*.

235. As an example, we will assume that the following table is given of a polynomial which is of the third degree in x, and of the third degree in y.

x	$y = 0$	2	3	5
2	9	65	159	599
3	19	133	289	979
5	51	401	741	2051
6	73	625	1099	2803

We first form the difference-table in x

9	65	159	599
10	68	130	380
2	22	32	52
0	2	3	5

and from this the difference-table in x and y

9	28	22	4
10	29	11	2
2	10	0	0
0	1	0	0

The polynomial $f(x\,;y)$ is completely determined by the given function-table; by means of the difference-table in x and y, and by formula (5), we may immediately write down

$$\begin{aligned} f(x\,;\,y) = {} & 9 + 10\,(x-2) + 2\,(x-2)\,(x-3) + 0 \\ & + y\,[28 + 29\,(x-2) + 10\,(x-2)\,(x-3) + (x-2)\,(x-3)\,(x-5)] \\ & + y\,(y-2)\,[22 + 11\,(x-2) + 0 + 0] \\ & + y\,(y-2)\,(y-3)\,[4 + 2\,(x-2) + 0 + 0] \end{aligned}$$

which reduces to

$$f(x;y) = x^3y + 2x^2 + xy^2 + 2xy^3 + 1.$$

236. If we only want the value of $f(x;\,y)$ at a special point, say $x = 4$, $y = 1$, we may avoid forming the difference-table in x and y, and content ourselves with the difference-table in x. By means of the latter, we form a table of $f(4;y)$ as shown below, and interpolate.

y					$f(4;y)$	f_1	f_2	f_3
0	$9+$	$10x_1+$	$2x_2+$	$0x_3=$	33			
						104		
2	65	68	22	2	241		44	
						236		8
3	159	130	32	3	477		84	
						488		
5	599	380	52	5	1453			

The result is

$$f(4;1) = 33 + 104\,y_1 + 44\,y_2 + 8\,y_3 = 109.$$

237. The same results are, of course, obtained by inserting the given values in Lagrange's formula (47).

238. If we interpolate by (44), we form, of course, only the differences actually used in the formula. Let the function-table be

x	$y=1$	2	3
2	1404	1989	2808
3	2340	3159	
4	3510		

As in this case the arguments are equidistant, we use ordinary, and not divided, differences, and form first the difference-table in x

1404	1989	2808
936	1170	
234		

and hence the difference-table in x and y

1404	585	234
936	234	
234		

The result of the interpolation is

$$f(x;y) = 1404 + [936\,(x-2) + 585\,(y-1)]$$

$$+\,234\left[\frac{(x-2)^{(2)}}{2} + (x-2)\,(y-1) + \frac{(y-1)^{(2)}}{2}\right] + R$$

or

$$f(x;y) = 117\,[(x+y)^2 + x - 2y + 3] + R,$$

R being given by (45) for $n = 2$. If it is known that $f(x; y)$ is a polynomial of the second degree in x and y *together*, that is, of the form

$$f(x;y) = \Sigma\, c_{\nu\mu}\, x^\nu\, y^\mu \qquad (\nu + \mu \leq 2),$$

we have evidently $R = 0$, and consequently

$$f(x;y) = 117\,[(x+y)^2 + x - 2y + 3].$$

239. As a slightly more complicated case, let us consider the following table of the function

$$f(x;y) = \frac{e^{xy} - 1}{x}$$

x	$y = 20$	22	24	26
0.035	28.9644	33.1362	37.6105	42.4092
0.040	30.6385	35.2725	40.2924	45.7304
0.045	32.4356	37.5830	43.2151	49.3776
0.050	34.3656	40.0833	46.4023	53.3859

and interpolate for $f(x; y)$ by the formula Bessel-Bessel with all the accuracy obtainable from the table. It is, then, (23) that is to be applied.

We first form the difference-table in x

31.5370	36.4278	41.7538	47.5540
1.7971	2.3105	2.9227	3.6472
0.1280	0.1820	0.2526	0.3436
0.0099	0.0156	0.0237	0.0351

The first column contains the central differences, or means, obtainable from the first column of the function-table, that is

$$31.5370 = \frac{1}{2}\,(30.6385 + 32.4356),\ 1.7971 = 32.4356 - 30.6385,$$

etc. The second column of the difference-table in x contains the corresponding values, formed by means of the second column of the function-table, and so on.

From the difference-table in x we now form, line by line, but otherwise by the same method, the difference-table in x and y, or

39.0908	5.3260	0.4547	0.0390
2.6166	0.6122	0.1056	0.0135
0.2173	0.0706	0.0185	0.0038
0.0197	0.0081	0.0029	0.0009

Here we have, then,

$$39.0908 = \frac{1}{2}(36.4278 + 41.7538),\ 5.3260 = 41.7538 - 36.4278,$$

and so on.

We may now write down the result, which is

$$f\left(x + \frac{1}{2}; y + \frac{1}{2}\right) =$$

$$39.0908 + 2.6166x + 0.2173\frac{x^2 - \frac{1}{4}}{2} + 0.0197\frac{x(x^2 - \frac{1}{4})}{6}$$

$$+ y\left(5.3260 + 0.6122x + 0.0706\frac{x^2 - \frac{1}{4}}{2} + 0.0081\frac{x(x^2 - \frac{1}{4})}{6}\right)$$

$$+ \frac{y^2 - \frac{1}{4}}{2}\left(0.4547 + 0.1056x + 0.0185\frac{x^2 - \frac{1}{4}}{2} + 0.0029\frac{x(x^2 - \frac{1}{4})}{6}\right)$$

$$+ \frac{y(y^2 - \frac{1}{4})}{6}\left(0.0390 + 0.0135x + 0.0038\frac{x^2 - \frac{1}{4}}{2} + 0.0009\frac{x(x^2 - \frac{1}{4})}{6}\right) + R,$$

where R is given by (24). The formula assumes as usual that the table interval is unity; it should therefore be noted that x and y have different meanings in the formula and in the function-table.

240. If we only want the value of $f(x; y) = \dfrac{e^{xy} - 1}{x}$ at a special point, say $x = 0.044$, $y = 22.5$, we proceed as in No. 236, stopping at the difference-table in x, and completing the calculation in the corresponding way as shown below, putting in the interpolation-formula

$$x = \frac{4}{5} - \frac{1}{2} = \frac{3}{10}, \; y = \frac{1}{4} - \frac{1}{2} = -\frac{1}{4}.$$

y						$f(.044;y)$	δ	δ^2	δ^3
20	31.5370 +	1.7971x +	0.1280$\frac{x^2-\frac{1}{4}}{2}$ +	0.0099$\frac{x(x^2-\frac{1}{4})}{6}$	=	32.0658			
							5.0405		
								0.4634	
22	36.4278	2.3105	0.1820	0.0156		37.1063			0.0429
							5.5039		
24	41.7538	2.9227	0.2526	0.0237		42.6102		0.5063	
							6.0102		
26	47 5540	3.6472	0.3436	0.0351		48.6204			

The result of the interpolation is

$f(0.044;\ 22.5) =$

$$39.8582 + 5.5039y + 0.4848\,\frac{y^2 - \frac{1}{4}}{2} + 0.0429\,\frac{y(y^2 - \frac{1}{4})}{6} = 38.4371.$$

The interpolation-formula for a single variable is, in practice, often written in the form §5 (18), see the numerical example in No. 45.

241. The question is now as to how many decimals can be relied upon in this result.

We observe first, that from $f(x\,;y) = \dfrac{e^{xy}-1}{x}$ we immediately obtain

$$D_y^4 = x^3e^{xy};\ D_x^4\,D_y^4 = ye^{xy}(x^3y^3 + 12x^2y^2 + 36xy + 24).$$

Further, we have

$$f(x;y) = \int^y e^{xt}dt,$$

and hence

$$D_x^4 = \int_0^y t^4\,e^{xt}dt < e^{xy}\int_0^y t^4dt = \tfrac{1}{5}\,y^5e^{xy}.$$

We now obtain upper limits to D_x^4, D_y^4 and $D_x^4\,D_y^4$ which are all positive in the intervals under consideration, by putting $x = 0.050$, $y = 26$, whence $xy = 1.30$. We find

$$D_x^4 < 8720000,\ D_y^4 < 0.00046,\ D_x^4\,D_y^4 < 8900.$$

But in order to obtain limits to the differential-coefficients in (24), we must take into account that the table intervals are not unity, and therefore multiply D_x^4 by 0.005^4, D_y^4 by $2^4 = 16$ and $D_x^4 D_y^4$ by 16×0.005^4. We thus find that the terms depending on D_ξ^4 or D_η^4 alone do not exceed 0.00008 and 0.00013 respectively, while the term depending on $D_\xi^4 D_\eta^4$ is quite negligible. The error represented by the remainder-term can therefore not exceed 2.1 units of the fourth decimal, to which may be added a trifling forcing-error. As it happens, all four decimals in the interpolated value are correct.

242. Interpolation with several variables is always a tedious process in comparison with interpolation with one variable, and it is therefore desirable, when possible, to reduce the problem to interpolation with one variable, as may sometimes be done. In the last example we may put

$$f(x;y) = y \cdot \frac{e^{xy} - 1}{xy} = y\,\psi\,(xy);$$

we then obtain from the function-table, after division by y, a table of ψ (z) in which we may interpolate for $\psi(0.99)$, whereafter the required value is found as $22.5\psi(0.99)$. In the special numerical case considered above the interpolation may even be completely avoided, as $0.045 \times 22 = 0.99$, so that $\psi(0.99) = \dfrac{37.5830}{22}$ and the required value $22.5\,\dfrac{37.5830}{22} = 38.4371$.

Not infrequently it is possible to interpolate along a single line of tabular values, for instance a diagonal or other sloping line.[1] Let us assume that the values of the function are known for integral (positive or negative) values of x and y. If we put

$$\psi(z) = f(kz + h; mz + n),$$

k, h, m and n being integers, $\psi(z)$ is known for integral values of z, and we have a table of $\psi(z)$ in which we may interpolate for non-integral values of z, and therefore find $f(x;y)$, if x and y have the form $x = kz + h$, $y = mz + n$.

[1] J. Spencer: Journal of the Institute of Actuaries, vol. 40, p. 299; W. Palin Elderton: Biometrica, vol. 6, p. 94.

§20. Mechanical Cubature

243. We define a formula of mechanical cubature as a formula, representing the approximate value of a *double integral*[1] as a linear function of a certain number of values of the function under consideration. We have therefore to do with an extension to two variables of the formulas derived in §16. If we confine ourselves to constant limits of integration, the method is quite analogous, as we evidently obtain a formula of the desired nature by integrating Lagrange's interpolation-formula for two variables, between fixed limits for x and y respectively. We may thus, as will appear presently, largely make use of the results obtained in §16. We content ourselves with a few applications of the quadrature-formulas involving an odd number of terms.

If we put

$$\left.\begin{aligned} P(x) &= x(x^2-1)\ldots(x^2-r^2), \quad P_\nu(x) = \frac{P(x)}{x-\nu}, \\ \overline{P}(y) &= y(y^2-1)\ldots(y^2-s^2), \quad \overline{P}_\mu(y) = \frac{\overline{P}(y)}{y-\mu}, \end{aligned}\right\} \quad (1)$$

Lagrange's formula with two variables, or §19 (47) with the remainder-term §19 (7), may be written

$$\left.\begin{aligned} f(x;y) = \sum_{\nu=-r}^{r}\sum_{\mu=-s}^{s} \frac{P_\nu(x)\,\overline{P}_\mu(y)}{P_\nu(\nu)\,\overline{P}_\mu(\mu)} f(\nu;\mu) + P(x) f(x,0,\pm 1,\ldots \pm r; y) \\ + \overline{P}(y) f(x;y,0,\pm 1,\ldots \pm s) - P(x)\,\overline{P}(y) f(x,0,\pm 1,\ldots \pm r; y, 0, \pm 1, \ldots \pm s). \end{aligned}\right\} \quad (2)$$

If we put, for abbreviation,

$$U_\nu = \int_{-m}^{m} \frac{P_\nu(x)}{P_\nu(\nu)}\,dx, \quad \overline{U}_\mu = \int_{-n}^{n} \frac{\overline{P}_\mu(y)}{\overline{P}_\mu(\mu)}\,dy, \quad (3)$$

and integrate (2) with respect to x between the limits $\pm m$, and with respect to y between the limits $\pm n$, we obtain

$$\int_{-n}^{n}\int_{-m}^{m} f(x;y)\,dxdy = \sum_{\nu=-r}^{r}\sum_{\mu=-s}^{s} U_\nu\,\overline{U}_\mu f(\nu;\mu) + R, \quad (4)$$

[1] This notion is often confounded with that of a *repeated integral*, see No. 117, 129 and 154. In the latter case we have only one variable, in the former two.

$$\left.\begin{aligned} R = \int_{-n}^{n}\int_{-m}^{m} [P(x)\,f(x, 0, \pm 1, \ldots \pm r; y) + \overline{P}\,(y)\,f(x; y, 0, \pm 1, \ldots \pm s) \\ - P(x)\,\overline{P}(y)\,f(x, 0, \pm 1, \ldots \pm r; y, 0, \pm 1, \ldots \pm s)]\,dxdy. \end{aligned}\right\}(5)$$

We may simplify this remainder-term by means of the properties of $P(x)$, proved in No. 169, and applying, *mutatis mutandis*, to $\overline{P}(y)$. We find for each of the three components of the remainder-term

$$\int_{-n}^{n}\int_{-m}^{m} P(x)\,f(x, 0, \pm 1, \ldots \pm r; y)\,dxdy$$

$$= - \int_{-n}^{n}\int_{-m}^{m} Q(x)\,f(x, x, 0, \pm 1, \ldots \pm r; y)\,dxdy$$

$$= - \frac{D_\xi^{2r+2} f(\xi; \eta)}{(2r+2)\,!}\int_{-n}^{n}\int_{-m}^{m} Q(x)\,dxdy = \frac{4n\,D_\xi^{2r+2} f(\xi; \eta)}{(2r+2)\,!}\int_{0}^{m} x^{[2r+2]}\,dx$$

and in the same way, or by substitution of letters,

$$\int_{-n}^{n}\int_{-m}^{m} \overline{P}(y)\,f(x; y, 0, \pm 1, \ldots \pm s)\,dxdy = \frac{4m\,D_\eta^{2s+2} f(\xi; \eta)}{(2s+2)\,!}\int_{0}^{n} y^{[2s+2]}\,dy;$$

finally

$$\int_{-n}^{n}\int_{-m}^{m} P(x)\,\overline{P}(y)\,f(x, 0, \pm 1, \ldots \pm r; y, 0, \pm 1, \ldots \pm s)\,dxdy$$

$$= - \int_{-n}^{n}\int_{-m}^{m} Q(x)\,\overline{P}(y)\,f(x, x, 0, \pm 1, \ldots \pm r; y, 0, \pm 1, \ldots \pm s)\,dxdy$$

$$= \int_{-n}^{n}\int_{-m}^{m} Q(x)\,\overline{Q}(y)\,f(x, x, 0, \pm 1, \ldots \pm r; y, y, 0, \pm 1, \ldots \pm s)\,dxdy$$

$$= \frac{D_\xi^{2r+2}\,D_\eta^{2s+2} f(\xi; \eta)}{(2r+2)\,!\,(2s+2)\,!}\int_{-m}^{m} Q(x)\,dx \,.\int_{-n}^{n} \overline{Q}(y)\,dy$$

$$= \frac{4\,D_\xi^{2r+2}\,D_\eta^{2s+2} f(\xi; \eta)}{(2r+2)!\;(2s+2)!}\int_{0}^{m} x^{[2r+2]}\,dx \,.\int_{0}^{n} y^{[2s+2]}\,dy,$$

so that

$$\left.\begin{aligned} R = \frac{4n\,D_\xi^{2r+2} f(\xi;\eta)}{(2r+2)\,!}\int_0^m x^{[2r+2]}\,dx + \frac{4m\,D_\eta^{2s+2} f(\xi;\eta)}{(2s+2)\,!}\int_0^n y^{[2s+2]}\,dy \\ - \frac{4\,D_\xi^{2r+2}\,D_\eta^{2s+2}\,f(\xi;\eta)}{(2r+2)!\,(2s+2)!}\int_0^m x^{[2r+2]}\,dx\,.\int_0^n y^{[2s+2]}\,dy. \end{aligned}\right\}(6)$$

244. If, for both the variables, we choose the interval of integration as unity, we must put, in (4),

$$f(x;y) = F\left(\frac{x}{2m};\frac{y}{2n}\right). \tag{7}$$

Introducing at the same time the notations

$$\left.\begin{aligned} V_\nu = \frac{1}{2m}\,U_\nu,\ \overline{V}_\mu = \frac{1}{2n}\,\overline{U}_\mu; \\ F_{\pm\nu 0} = F\left(\frac{\nu}{2m};0\right) + F\left(-\frac{\nu}{2m};0\right),\ F_{\pm 0\mu} = F\left(0;\frac{\mu}{2n}\right) + F\left(0;-\frac{\mu}{2n}\right), \\ F_{\pm\nu\mu} = F\left(\frac{\nu}{2m};\frac{\mu}{2n}\right) + F\left(\frac{\nu}{2m};-\frac{\mu}{2n}\right) + F\left(-\frac{\nu}{2m};\frac{\mu}{2n}\right) + F\left(-\frac{\nu}{2m};-\frac{\mu}{2n}\right), \end{aligned}\right\}(8)$$

and writing, as in §16 (14),

$$O_{2m}^{[2k]} = \frac{2}{(2k)\,!\,(2m)^{2k+1}}\int_0^m x^{[2k]}\,dx, \tag{9}$$

(4) and (6) may, since according to No. 168

$$U_{-\nu} = U_\nu,\ \overline{U}_{-\mu} = \overline{U}_\mu,$$

be written

$$\left.\begin{aligned} \int_{-\frac12}^{\frac12}\int_{-\frac12}^{\frac12} F(x;y)dxdy = V_0\overline{V}_0 F_{00} + \overline{V}_0\sum_1^r V_\nu F_{\pm\nu 0} + V_0\sum_1^s \overline{V}_\mu F_{\pm 0\mu} \\ + \sum_{\nu=1}^r\sum_{\mu=1}^s V_\nu\overline{V}_\mu F_{\pm\nu\mu} + R, \end{aligned}\right\}(10)$$

$$R = O_{2m}^{[2r+2]} D_\xi^{2r+2} + O_{2n}^{[2s+2]} D_\eta^{2s+2} - O_{2m}^{[2r+2]} O_{2n}^{[2s+2]} D_\xi^{2r+2} D_\eta^{2s+2} \tag{11}$$

In the two most important cases, $r = m$, $s = n$ (the closed type), and $r = m - 1$, $s = n - 1$ (the open type), the coefficients V_ν, $\overline{V}_\mu$, $O_{2m}^{[2r+2]}$ and $O_{2n}^{[2s+2]}$ may be immediately taken from the tables in No. 172.

It should be noted that the right-hand side of (10), apart from the remainder-term, is the symbolical product of two formulas with one argument, viz.

$$\left.\begin{aligned} &V_0 + \sum_1^r V_\nu \left(E_u^\nu + E_u^{-\nu}\right) \\ \text{and} \\ &\overline{V}_0 + \sum_1^s \overline{V}_\mu \left(E_v^\mu + E_v^{-\mu}\right) \end{aligned}\right\} \quad (12)$$

the factor left out being $F\left(\frac{u}{2m}; \frac{v}{2n}\right)$.

245. The most important of these formulas is *Simpson's formula for two variables* which is obtained for $r = m = 1$, $s = n = 1$, that is

$$\int_{-\frac{1}{2}}^{\frac{1}{2}}\int_{-\frac{1}{2}}^{\frac{1}{2}} F(x;y)\,dxdy = \frac{16\,F_{00} + 4\,(F_{\pm 10} + F_{\pm 01}) + F_{\pm 11}}{36} + R, \quad (13)$$

$$R = \frac{-1}{2880}\left(D_\xi^4 + D_\eta^4 + \frac{1}{2880}\,D_\xi^4\,D_\eta^4\right). \quad (14)$$

This formula contains 9 values of $F(x;y)$, placed in a square. F_{00} is the value in the middle of the square, $F_{\pm 11}$ is the sum of the values at the corners, and $(F_{\pm 10} + F_{\pm 01})$ the sum of the remaining values.

246. If, next, we put $r = m = 2$, $s = n = 2$, we get a formula of considerably greater accuracy, viz.

$$\left.\begin{aligned} \int_{-\frac{1}{2}}^{\frac{1}{2}}\int_{-\frac{1}{2}}^{\frac{1}{2}} F(x;y)dxdy &= \frac{1}{8100}\,[144\,F_{00} + 1024\,F_{\pm 11} + 49\,F_{\pm 22} \\ + 384\,(F_{\pm 01} + F_{\pm 10}) &+ 84\,(F_{\pm 02} + F_{\pm 20}) + 224\,(F_{\pm 12} + F_{\pm 21})] + R \end{aligned}\right\} \quad (15)$$

$$R = -\,0.0^652\,(D_\xi^6 + D_\eta^6 + 0.0^652\,D_\xi^6\,D_\eta^6). \quad (16)$$

In this formula we have used 25 values, placed in a square schedule as follows:

F_{00}	The middle.
$F_{\pm 11}$	The corners of the inner square.
$F_{\pm 22}$	The corners of the outer square.
$(F_{\pm 01} + F_{\pm 10})$	The middles of the sides of the inner square.
$(F_{\pm 02} + F_{\pm 20})$	The middles of the sides of the outer square.
$(F_{\pm 12} + F_{\pm 21})$	The remaining 8 points.

In performing the calculation, it is practical to mark off the values as they are used.

247. In cases where only a rough approximation is required, we may sometimes replace the volume whose base is a square with side 1, by the height in the middle of the square. That is, we apply the simplest formula of the open type, or §16 (15). The result is

$$\int_{-\frac{1}{2}}^{\frac{1}{2}} \int_{-\frac{1}{2}}^{\frac{1}{2}} F(x;y)\, dxdy = F_{00} + \frac{1}{24}\left(D_\xi^2 + D_\eta^2 - \frac{1}{24} D_\xi^2 D_\eta^2\right). \quad (17)$$

248. As an application, let us calculate the integral

$$\int_1^3 \int_2^3 \frac{x}{x^2+y^2}\, dxdy. \quad (18)$$

We have then

$$F(x;y) = \frac{x}{x^2+y^2} = \frac{1}{2}\left(\frac{1}{x+iy} + \frac{1}{x-iy}\right),$$

so that

$$D_x^{2\nu} D_y^{2\mu} = (-1)^\mu \frac{(2\nu+2\mu)!}{2}\left[\frac{1}{(x+iy)^{2\nu+2\mu+1}} + \frac{1}{(x-iy)^{2\nu+2\mu+1}}\right].$$

If on the right we put

$$x = \rho \cos \theta, \quad y = \rho \sin \theta,$$

whence

$$x + iy = \rho e^{i\theta}, \quad x - iy = \rho e^{-i\theta},$$

we get

$$D_x^{2\nu} D_y^{2\mu} = (-1)^\mu \frac{(2\nu + 2\mu)!}{\rho^{2\nu+2\mu+1}} \cos (2\nu + 2\mu + 1)\theta,$$

consequently

$$\left| D_x^{2\nu} D_y^{2\mu} \right| \leq \frac{(2\nu + 2\mu)!}{\rho^{2\nu+2\mu+1}}.$$

For the region of integration under consideration we may, on the right, put $x = y = 2$, and thus find

$$\left| D_\xi^{2\nu} D_\eta^{2\mu} \right| \leq \frac{(2\nu + 2\mu)!}{2^{3\nu+3\mu+1}\sqrt{2}} \tag{19}$$

to be used in examining the remainder-term.

If we apply Simpson's formula for two variables, the function-table has the following appearance

x	$y = 2$	$2\frac{1}{2}$	3
2	$\frac{1}{4}$	$\frac{8}{41}$	$\frac{2}{13}$
$2\frac{1}{2}$	$\frac{10}{41}$	$\frac{1}{5}$	$\frac{10}{61}$
3	$\frac{3}{13}$	$\frac{12}{61}$	$\frac{1}{6}$

so that the required value, apart from the remainder-term, is

$$\frac{1}{36}\left[\frac{16}{5} + 4\left(\frac{18}{41} + \frac{22}{61}\right) + \frac{1}{4} + \frac{5}{13} + \frac{1}{6}\right] = \frac{14045621}{70228080} = 0.200000.$$

The question is now as to how many of these figures may be relied upon.

We find by (19) and (14)

$$|R| < \frac{1}{2880}\left(\frac{4!}{2^7\sqrt{2}} + \frac{4!}{2^7\sqrt{2}} + \frac{1}{2880}\frac{8!}{2^{13}\sqrt{2}}\right) < 0.000094,$$

so that the required volume is 0.2000 with an error that does not exceed one unit of the fourth decimal.

In the case under consideration it is, by the way, possible to perform the integration rigourously; we thus find the value

$$\frac{9}{2}\text{Log } 2 + 3 \text{ Log } 3 - \frac{5}{2}\text{Log } 13 + \frac{\pi}{4} - \text{arctg}\frac{2}{3} = 0.200021,$$

so that the limit to the error, found by the remainder-term, is not much too high.

249. We will now treat the same example by formula (15). The function-table is here

x	y = 2.00	2.25	2.50	2.75	3.00
2.00	0.25	0.2206896	0.1951220	0.1729730	0.1538462
2.25	0.2482759	0.2222222	0.1988951	0.1782178	0.16
2.50	0.2439024	0.2209945	0.2	0.1809955	0.1639344
2.75	0.2378379	0.2178218	0.1990950	0.1818182	0.1660377
3.00	0.2307693	0.2133333	0.1967213	0.1811321	0.1666667

and the required value, by (15),

$$\frac{1}{8100}(144 \times 0.2 + 1024 \times 0.8000800 + 49 \times 0.8012822$$

$$+384 \times 0.7999801 + 84 \times 0.7996801 + 224 \times 1.6002795) = 0.200021.$$

For the remainder-term we find by (16) and (19)

$$|R| < 0.0^652\left(\frac{6!}{2^{10}\sqrt{2}} + \frac{6!}{2^{10}\sqrt{2}} + 0.0^652\frac{12!}{2^{19}\sqrt{2}}\right) < 0.0^652.$$

In this case we may be sure that the error can only slightly exceed one half unit of the sixth decimal; as a matter of fact, this is also correct.

250. If, finally, we apply (17), we find for the double integral the value 0.2, and as a limit to the error, by (19)

$$|R| < \frac{1}{24}\left(\frac{2}{2^4\sqrt{2}} + \frac{2}{2^4\sqrt{2}} + \frac{1}{24}\frac{4!}{2^7\sqrt{2}}\right) < 0.0077.$$

The result should, therefore, be stated as 0.20 with an error that does not exceed one unit of the second decimal. But comparison with the exact value shows that we have, as a matter of fact, obtained four correct decimals by the rough method. This example may, therefore, be taken as a useful warning against the too often employed method: namely, that of judging the value of an approximate formula by the results it has produced in isolated numerical cases, instead of examining the remainder-term.

APPENDIX

§21. On Differential Coefficients of Arbitrary Order

251. The examination of the error involved in an interpolation-formula is, in the general case, based on the knowledge of the differential coefficient of a certain order. The direct calculation of the latter is as a rule possible, but may occasionally lead to tedious calculations, even in the case of elementary functions. Although the investigation of these questions does not really belong to the theory of interpolation, but is a subject of mathematical analysis, yet we find it practical to collect here some of the expressions known for differential coefficients of arbitrary order, and to call attention to the means we possess of calculating or estimating such differential coefficients.

252. We assume that the reader is familiar with the result of differentiating one of the elementary functions

$$x^a, \operatorname{Log} x, a^x, \sin x, \cos x$$

n times; we also assume Leibnitz' formula for the nth differential coefficient of the product of two functions u and v

$$D^n uv = \sum_{\nu=0}^{n} \binom{n}{\nu} D^\nu u \cdot D^{n-\nu} v. \tag{1}$$

We remind the reader, *en passant,* of the fact that any formula for the nth differential coefficient can be generalized by a linear transformation of the variable, as

$$D^n f(a + bx) = b^n f^{(n)}(a + bx). \tag{2}$$

253. Every rational function of x may, as is well known, be written as a sum of a polynomial and a finite number of terms of the form $\dfrac{a}{(x-b)^m}$, where m is a positive integer, while a and b may be any constants, real or complex. We may therefore find an explicit expression for the nth differential coefficient. The result may finally be brought into real form; for our purpose most profitably by introducing polar coordinates, as in the example of No. 248.

In order to apply this method with advantage it is, however, necessary that the polynomial in the denominator is, or can easily be, resolved into factors. In the case where all the roots are different, the result is particularly simple. Let us put

$$P(x) = (x - a_1)(x - a_2) \ldots (x - a_p), \tag{3}$$

and let $F(x)$ denote a polynomial which is of *lower* degree than $P(x)$. We have then

$$\frac{F(x)}{P(x)} = \sum_{\nu=1}^{p} \frac{F(a_\nu)}{P'(a_\nu)} \frac{1}{x - a_\nu}; \tag{4}$$

for if we multiply on both sides by $P(x)$, the right-hand side will evidently represent a polynomial which is, at the most, of degree $p - 1$, and which, for $x = a_\nu$ $(\nu = 1, 2, \ldots p)$, assumes the values $F(a_\nu)$ $(\nu = 1, 2, \ldots p)$. This polynomial must, therefore, be identical with $F(x)$, as the degree of the latter does not exceed $p - 1$.

$P'(a_\nu)$ may either be obtained by differentiation of $P(x)$, being thus expressed in terms of the coefficients of the polynomial, or be expressed in terms of the roots as

$$\left.\begin{aligned} P'(a_\nu) &= \lim_{x \to a_\nu} \frac{P(x)}{x - a_\nu} \\ &= (a_\nu - a_1) \ldots (a_\nu - a_{\nu-1}) . (a_\nu - a_{\nu+1}) \ldots (a_\nu - a_p). \end{aligned}\right\} \tag{5}$$

We now obtain from (4) the desired formula

$$D^n \frac{F(x)}{P(x)} = (-1)^n n! \sum_{\nu=1}^{p} \frac{F(a_\nu)}{P'(a_\nu)} \frac{1}{(x - a_\nu)^{n+1}}. \tag{6}$$

254. As an example, we will treat the case where $P(x)$ is of the second, $F(x)$ of the first degree.

If $P(x)$ has the real and different roots a and b, we find from (6)

$$D^n \frac{\alpha + \beta x}{(x-a)(x-b)} = \frac{(-1)^n n!}{a-b}\left[\frac{\alpha + \beta a}{(x-a)^{n+1}} - \frac{\alpha + \beta b}{(x-b)^{n+1}}\right]. \quad (7)$$

If the roots are real and equal, we have

$$\frac{\alpha + \beta x}{(x-a)^2} = \frac{\beta}{x-a} + \frac{\alpha + \beta a}{(x-a)^2},$$

consequently

$$D^n \frac{\alpha + \beta x}{(x-a)^2} = \frac{(-1)^n n!}{(x-a)^{n+2}}[\beta(x-a) + (n+1)(\alpha + \beta a)]. \quad (8)$$

If finally—which is the most interesting case—the roots are complex $a \pm ib$, we find by (6)

$$D^n \frac{\alpha + \beta x}{(x-a)^2+b^2} = \frac{(-1)^n n!}{2ib}\left[\frac{\alpha+\beta(a+ib)}{(x-a-ib)^{n+1}} - \frac{\alpha+\beta(a-ib)}{(x-a+ib)^{n+1}}\right]. \quad (9)$$

If, in this, we put

$$\left.\begin{array}{ll} \left.\begin{array}{l} x - a + ib = \rho e^{i\theta} \\ x - a - ib = \rho e^{-i\theta} \end{array}\right\} & \begin{array}{l} \rho = \sqrt{(x-a)^2 + b^2} \\ \theta = \operatorname{arctg} \dfrac{b}{x-a} \end{array} \end{array}\right\} \quad (10)$$

we obtain the real form

$$D^n \frac{\alpha+\beta x}{(x-a)^2+b^2} = \frac{(-1)^n n!}{b\rho^{n+1}}[\beta b \cos(n+1)\theta + (\alpha+\beta a)\sin(n+1)\theta]. \quad (11)$$

Let us, in particular, put

$$\alpha = 1, \beta = 0, a = 0, b = 1;$$

remembering that

$$\frac{1}{1+x^2} = D \operatorname{arctg} x$$

we find, for $n > 0$,

$$D^n \operatorname{arctg} x = (-1)^{n-1} \frac{(n-1)! \sin(n \operatorname{arc\,cot} x)}{(1+x^2)^{\frac{n}{2}}}. \quad (12)$$

If, on the other hand, we put

$$\alpha = 0, \beta = 1, a = 0, b = 1$$

and remember that

$$\frac{2x}{1+x^2} = D \operatorname{Log}(1+x^2),$$

we find, for $n > 0$,

$$D^n \operatorname{Log}(1+x^2) = (-1)^{n-1}\frac{2(n-1)!\cos(n \operatorname{arc} \cot x)}{(1+x^2)^{\frac{n}{2}}}. \tag{13}$$

We may evidently also express (9) in algebraical-real form, but this is not advantageous for our purpose.

255. We next recall the fact that in any case where we are able to expand $f(x+h)$ in powers of h, we obtain at the same time an expression for $D^n f(x)$, as the coefficient of h^n is $\frac{1}{n!} D^n f(x)$. This is also so, if the differential coefficients only exist as far as the order used in the remainder-term; it is, however, convenient to write ∞ as the upper limit of summation, although we do not in reality assume that the expansion can be continued indefinitely.

These remarks lead in certain cases, important practically, to simple expressions for $D^n f(y_x)$, where y_x is a function of x. The method is as follows. If $f(y_{x+h})$ can be expanded in powers of h, the coefficient of h^n is $\frac{1}{n!} D^n f(y_x)$, where D acts on x. We now put

$$\begin{aligned} f(y_{x+h}) &= f[y_x + (y_{x+h} - y_x)] \\ &= \sum_0^\infty \frac{1}{n!} f^{(n)}(y_x)(y_{x+h} - y_x)^n \end{aligned}$$

$$= \sum_{n=0}^\infty \frac{1}{n!} f^{(n)}(y_x)\left(A_0^{(n)} h^n + A_1^{(n)} h^{n+1} + A_2^{(n)} h^{n+2} + \ldots\right),$$

so that

$$\frac{1}{n!} D^n f(y_x) = \frac{A_0^{(n)}}{n!} f^{(n)}(y_x) + \frac{A_1^{(n-1)}}{(n-1)!} f^{(n-1)}(y_x) + \frac{A_2^{(n-2)}}{(n-2)!} f^{(n-2)}(y_x) + \ldots$$

The result may therefore be written

$$\left.\begin{aligned} D^n f(y_x) &= \sum_{\nu=0}^{n} A_\nu^{(n-\nu)} n^{(\nu)} f^{(n-\nu)}(y_x), \\ (y_{x+h} - y_x)^r &= h^r \sum_{\nu=0}^{\infty} A_\nu^{(r)} h^\nu. \end{aligned}\right\} \quad (14)$$

256. This formula leads, for instance, to the desired result, if y_x has one of the following four forms, important in practice

$$a + bx + cx^2, \quad \frac{ax+b}{cx+d}, \quad e^{a+bx}, \quad \operatorname{Log}(a+bx), \qquad (15)$$

which we will treat in succession.

If $y_x = a + bx + cx^2$, we have

$$(y_{x+h} - y_x)^r = h^r (b + 2cx + ch)^r.$$

Expanding the right-hand side by the binomial theorem, we find

$$A_\nu^{(r)} = \binom{r}{\nu} c^\nu (b + 2cx)^{r-\nu},$$

so that the result is

$$\left.\begin{aligned} D^n f(y_x) &= \sum_{\nu=0}^{\leq \frac{n}{2}} \frac{n^{(2\nu)}}{\nu!} c^\nu (b + 2cx)^{n-2\nu} f^{(n-\nu)}(y_x), \\ y_x &= a + bx + cx^2. \end{aligned}\right\} \quad (16)$$

If, in particular, $a = b = 0$, $c = 1$, we get

$$D^n f(x^2) = \sum_{\nu=0}^{\leq \frac{n}{2}} \frac{n^{(2\nu)}}{\nu!} (2x)^{n-2\nu} f^{(n-\nu)}(x^2), \qquad (17)$$

and as a special case of this formula

$$\left.\begin{aligned} D^n e^{-\frac{x^2}{2}} &= e^{-\frac{x^2}{2}} H_n(x), \\ H_n(x) &= \sum_{\nu=0}^{\leq \frac{n}{2}} (-1)^{n-\nu} \frac{n^{(2\nu)}}{\nu!\, 2^\nu} x^{n-2\nu}. \end{aligned}\right\} \quad (18)$$

The polynomials $H_n(x)$ are usually called Hermite's polynominals[1] although they were first found by Tchebychef.

Another application is

$$D^n \frac{1}{\sqrt{1-x^2}} = \frac{1}{\sqrt{1-x^2}} \sum_{\nu=0}^{\leq \frac{n}{2}} \frac{n^{(2\nu)}.1.3. \ldots (2n-2\nu-1)}{\nu!\, 2^\nu} \frac{x^{n-2\nu}}{(1-x^2)^{n-\nu}}, \qquad (19)$$

from which we obtain, as

$$\frac{1}{\sqrt{1-x^2}} = D \operatorname{arc} \sin x,$$

the following formula, valid for $n > 0$,

$$D^n \operatorname{arc} \sin x = \frac{1}{\sqrt{1-x^2}} \sum_{\nu=0}^{\leq \frac{n-1}{2}} \frac{(n-1)^{(2\nu)}.1.3. \ldots (2n-2\nu-3)}{\nu!\, 2^\nu} \frac{x^{n-2\nu-1}}{(1-x^2)^{n-\nu-1}}. \qquad (20)$$

257. If, next, we put

$$y_x = \frac{ax+b}{cx+d},$$

we have

$$(y_{x+h} - y_x)^r = h^r \left(1 + \frac{ch}{cx+d}\right)^{-r} \frac{(ad-cb)^r}{(cx+d)^{2r}},$$

so that

$$A_\nu^{(r)} = (-1)^\nu \binom{r+\nu-1}{\nu} \left(\frac{c}{cx+d}\right)^\nu \frac{(ad-cb)^r}{(cx+d)^{2r}}.$$

We therefore obtain the result in the following form

$$\left.\begin{aligned} D^n f(y_x) &= \frac{(ad-cb)^n}{(cx+d)^{2n}} \sum_{\nu=0}^{n-1} (-1)^\nu \binom{n}{\nu} (n-1)^{(\nu)} \left(c\,\frac{cx+d}{ad-cb}\right)^\nu f^{(n-\nu)}(y_x) \\ y_x &= \frac{ax+b}{cx+d}. \end{aligned}\right\} \qquad (21)$$

[1] A table of these polynomials as far as $n = 6$ is found in N. R. Jörgensen: Undersögelser over Frequensflader og Korrelation, Copenhagen 1916.

In particular, we find for $b = c = 1$, $a = d = 0$,

$$D^n f\left(\frac{1}{x}\right) = \frac{(-1)^n}{x^{2n}} \sum_{\nu=0}^{n-1} \binom{n}{\nu} (n-1)^{(\nu)} x^\nu f^{(n-\nu)}\left(\frac{1}{x}\right) \tag{22}$$

whence

$$D^n e^{\frac{1}{x}} = \frac{(-1)^n}{x^{2n}} e^{\frac{1}{x}} \sum_{\nu=0}^{n-1} \binom{n}{\nu} (n-1)^{(\nu)} x^\nu. \tag{23}$$

258. If $y_x = e^{a+bx}$, we have

$$(y_{x+h} - y_x)^r = e^{r(a+bx)}\,(e^{bh} - 1)^r = e^{r(a+bx)} \sum_{s=r}^{\infty} \frac{\Delta^r 0^s}{s!} b^s h^s,$$

where we have first developed $(e^{bh} - 1)^r$ in powers of e^{bh} and then in powers of h. Therefore

$$A_\nu^{(r)} = e^{r(a+bx)} \frac{\Delta^r 0^{r+\nu}}{(r+\nu)!} b^{r+\nu}$$

and finally, for $n > 0$,

$$\left.\begin{aligned} D^n f(y_x) &= b^n \sum_{\nu=1}^{n} \frac{\Delta^\nu 0^n}{\nu!} e^{\nu(a+bx)} f^{(\nu)}(y_x), \\ y_x &= e^{a+bx}. \end{aligned}\right\} \tag{24}$$

From this we find, for instance,

$$D^n f(e^x) = \sum_{\nu=1}^{n} \frac{\Delta^\nu 0^n}{\nu!} e^{\nu x} f^{(\nu)}(e^x); \tag{25}$$

in particular

$$D^n \frac{1}{e^x - 1} = \frac{1}{e^x - 1} \sum_{\nu=1}^{n} (-1)^\nu \frac{\Delta^\nu 0^n}{(1 - e^{-x})^\nu}, \tag{26}$$

$$D^n e^{e^x} = e^{e^x} \sum_{\nu=1}^{n} \frac{\Delta^\nu 0^n}{\nu!} e^{\nu x}, \tag{27}$$

$$D^n \operatorname{Log}(e^x - 1) = \sum_{\nu=1}^{n} \frac{(-1)^{\nu+1}}{\nu} \frac{\Delta^\nu 0^n}{(1 - e^{-x})^\nu}. \tag{28}$$

259. Functions of $\sin x$ and $\cos x$ may be treated by (24). If, for instance, we write

$$\cot x = i\frac{e^{xi}+e^{-xi}}{e^{xi}-e^{-xi}} = i + \frac{2i}{e^{2xi}-1},$$

we find by (24), for $n > 0$,

$$D^n \cot x = D^n \frac{2i}{e^{2xi}-1} = \frac{(2i)^{n+1}}{e^{2xi}-1}\sum_{\nu=1}^{n}(-1)^{\nu}\frac{\Delta^{\nu}0^n}{(1-e^{-2xi})^{\nu}}$$

$$= \sum_{\nu=1}^{n}(-1)^{\nu}\frac{\Delta^{\nu}0^n}{\sin^{\nu+1}x}(2i)^{n-\nu}e^{(\nu-1)xi}.$$

If, in this, we put $i = e^{\frac{\pi i}{2}}$, we get

$$D^n \cot x = \sum_{\nu=1}^{n}(-1)^{\nu}\frac{2^{n-\nu}\Delta^{\nu}0^n}{\sin^{\nu+1}x}e^{[(\nu-1)x+(n-\nu)\frac{\pi}{2}]i};$$

but as $D^n\cot x$ is real for real x, the imaginary part must vanish identically so that we finally obtain

$$D^n \cot x = \sum_{\nu=1}^{n}(-1)^{\nu}2^{n-\nu}\Delta^{\nu}0^n\frac{\cos\left[(\nu-1)x+(n-\nu)\frac{\pi}{2}\right]}{\sin^{\nu+1}x}. \tag{29}$$

As

$$\cot\left(x+\frac{\pi}{2}\right) = -\operatorname{tg}x,$$

we find by (29)

$$D^n \operatorname{tg} x = \sum_{\nu=1}^{n}(-1)^{\nu+1}2^{n-\nu}\Delta^{\nu}0^n\frac{\sin\left[(\nu-1)x+\frac{n\pi}{2}\right]}{\cos^{\nu+1}x}. \tag{30}$$

It should further be noted that, as

$$\cot x = D\operatorname{Log}\sin x, \quad \operatorname{tg}x = -D\operatorname{Log}\cos x,$$

we obtain from (29) and (30) the following formulas, valid for $n > 1$,

$$D^n \operatorname{Log} \sin x = \sum_{\nu=1}^{n-1} (-1)^\nu 2^{n-\nu-1} \Delta^\nu 0^{n-1} \frac{\sin\left[(\nu-1)x+(n-\nu)\frac{\pi}{2}\right]}{\sin^{\nu+1} x}, \tag{31}$$

$$D^n \operatorname{Log} \cos x = \sum_{\nu=1}^{n-1} (-1)^{\nu+1} 2^{n-\nu-1} \Delta^\nu 0^{n-1} \frac{\cos\left[(\nu-1)x+\frac{n\pi}{2}\right]}{\cos^{\nu+1} x}. \tag{32}$$

From these two formulas we derive by subtraction general expressions for $D^n \operatorname{Log} \operatorname{tg} x$ and $D^n \operatorname{Log} \cot x$, and from these again general expressions for $D^n \operatorname{cosec} x$ and $D^n \sec x$, by noting that

$$\operatorname{cosec} x = D \operatorname{Log} \operatorname{tg} \frac{x}{2}, \quad \sec x = D \operatorname{Log} \cot\left(\frac{x}{2} - \frac{\pi}{4}\right).$$

260. If, finally, we put $y_x = \operatorname{Log}(a + bx)$, we have

$$(y_{x+h} - y_x)^r = \operatorname{Log}^r\left(1 + \frac{bh}{a+bx}\right) = D^r_{t=0}\left(1 + \frac{bh}{a+bx}\right)^t$$

$$= D^r_{t=0} \sum_{s=0}^{\infty} \frac{t^{(s)}}{s!}\left(\frac{b}{a+bx}\right)^s h^s = \sum_{s=r}^{\infty} \frac{D^r 0^{(s)}}{s!}\left(\frac{b}{a+bx}\right)^s h^s,$$

consequently

$$A_\nu^{(r)} = \frac{D^r 0^{(r+\nu)}}{(r+\nu)!}\left(\frac{b}{a+bx}\right)^{r+\nu}$$

and therefore, for $n > 0$,

$$\left.\begin{aligned} D^n f(y_x) &= \left(\frac{b}{a+bx}\right)^n \sum_{\nu=1}^{n} (-1)^{n+\nu} \frac{D^\nu 0^{(-n)}}{\nu!} f^{(\nu)}(y_x), \\ y_x &= \operatorname{Log}(a+bx). \end{aligned}\right\} \tag{33}$$

In particular, we have

$$D^n f(\operatorname{Log} x) = \frac{1}{x^n} \sum_{\nu=1}^{n} \frac{(-1)^{n+\nu}}{\nu!} D^\nu 0^{(-n)} f^{(\nu)}(\operatorname{Log} x), \tag{34}$$

whence, for instance,

$$D^n \frac{1}{\operatorname{Log} x} = \frac{(-1)^n}{x^n} \sum_{\nu=1}^{n} \frac{D^\nu 0^{(-n)}}{\operatorname{Log}^{\nu+1} x}. \tag{35}$$

As the Logarithmic-integral function is defined by[1]

$$\operatorname{Li}(x) = \lim_{h \to 0} \left(\int_0^{1-h} \frac{dt}{\operatorname{Log} t} + \int_{1+h}^{x} \frac{dt}{\operatorname{Log} t} \right),$$

we have

$$D \operatorname{Li}(x) = \frac{1}{\operatorname{Log} x}$$

and hence, by (35), for $n > 1$

$$D^n \operatorname{Li}(x) = \frac{(-1)^{n-1}}{x^{n-1}} \sum_{\nu=1}^{n-1} \frac{D^\nu 0^{(-n+1)}}{\operatorname{Log}^{\nu+1} x}. \tag{36}$$

261. Instead of deriving (16), (21), (24) and (33) directly, we might, of course, have derived the more special formulas (17), (22), (25) and (34) directly, and thereafter have obtained the more general formulas by a linear transformation of the variable.

262. In many cases where a function is defined by a series or by a definite integral we may, by differentiation of the series or the integral, obtain explicit expressions for the nth differential coefficient and limits to this. An important example is the function $\operatorname{Log} \Gamma(x)$. From the well-known expansion[2]

$$\operatorname{Log} \Gamma(x) = -Cx - \operatorname{Log} x + \sum_{1}^{\infty} \left[\frac{x}{\nu} - \operatorname{Log}\left(1 + \frac{x}{\nu}\right) \right] \tag{37}$$

we thus obtain

$$D^n \operatorname{Log} \Gamma(x) = (-1)^n (n-1)! \sum_{0}^{\infty} \frac{1}{(x+\nu)^n} \quad (n > 1). \tag{38}$$

[1] Landau: Handbuch der Lehre von der Verteilung der Primzahlen, vol. I, p. 27.

[2] See, for instance, Whittaker and Watson: Modern Analysis, third ed., p. 236.

Now we have, by §12 (29), assuming $x > 0$ and $0 < \theta < 1$,

$$\sum_{0}^{\infty} \frac{1}{(x+\nu)^n} = \frac{\theta}{x^n} + \int_0^{\infty} \frac{dt}{(x+t)^n} = \frac{\theta}{x^n} + \frac{1}{(n-1)x^{n-1}},$$

and consequently

$$|D^n \operatorname{Log} \Gamma(x)| = \frac{(n-2)!}{x^n}[x + \theta(n-1)] \quad \begin{pmatrix} x > 0, n > 1 \\ 0 < \theta < 1 \end{pmatrix}, \tag{39}$$

which is very useful in interpolating in a table of $\operatorname{Log} \Gamma(x)$.

It may also be noted, that $D^n \operatorname{Log} \Gamma(x)$, as appears from (38), does not change its sign, so that the Error-Test is applicable.

263. Mathematical analysis has still other means at its disposal for ascertaining limits to the differential coefficients of functions of one or more variables, especially the inequalities found by Cauchy[1] which only require that certain general analytical properties of the function are known. We could not, however, go into this subject without transgressing the scope of this book.

[1] See, for instance, Picard: Traité d'Analyse, vol. II, p. 111 and p. 238–240.

INDEX

Ascending difference, 7
Ascending factorial, 8
Associative property, 5

Bernoullian function, periodic, 130
Bernoulli's numbers, 120, 121, 129
Bernoulli's polynomials,
—— definition, 119, 128
—— recurrence-formula, 120
—— table of, 120
—— explicit expressions, 121, 125, 127
—— symbolical relations, 121–122
—— properties of, 123, 124, 128
—— Jacobi's theorem, 124–127
Bessel's interpolation-formula,
—— one variable, 30–31, 32, 40, 41, 43, 45, 64, 67–68, 72, 77, 109, 189
—— two variables, 210–212, 220–223
Biermann, O., 215
Boole, G., 180
Briggs, 72
Bromwich, 129

Cauchy, 241
Central difference, 7
Central differences of nothing, 55–57, 60
Central factorial, 8
Closed type of quadrature-formula, 158, 166
Coefficients, factorial, 53–60
Coefficients, see also Numbers.
Commutative property, 5
Constants, see Numbers.
Cotes' method of mechanical quadrature, 154
Cubature, see Mechanical cubature.

Derivative, see Differential coefficient.
Descending difference, 7
Descending factorial, 8
Difference-equations, 180–184
Differences,
—— descending, ascending, central, 7–10
—— divided, see Divided differences.
—— expressed by differential coefficients, 68–71, 198
—— of nothing, 54–57, 60
Difference-table,
—— descending, ascending, central differences, 7
—— divided differences, 14
—— repeated arguments, 21
—— in x and y, 217–221
Differential coefficients of arbitrary order,
—— generally,231, 234–235
—— of a product, 231
—— of a rational function, 231–233
—— of $f(a + bx + cx^2)$, 235–236
—— of $f\left(\frac{ax + b}{cx + d}\right)$, 236–237
—— of $f(e^{a + bx})$, 237
—— of $f(\sin x, \cos x)$, 238–239
—— of $f(\mathrm{Log}\ (a + bx))$, 239–240
—— of arc tg x, 233
—— of arc sin x, 236
——of $e^{-\frac{x^2}{2}}$, 235
—— of $\mathrm{Li}(x)$, 240
—— of $\mathrm{Log}\ \Gamma(x)$, 240–241
Differential coefficients of nothing, 57–60
Differential equations, numerical integration of, 170–177
Differentiation, numerical, see Numerical differentiation.
Dirichlet's integral-formula, 93

Dirichlet's sum-formula, 91–92
Displacement-symbol, 4–6
Distributive property, 5
Divided differences,
—— definition, 14
—— difference-table, 14
—— expressed by values of function, 15
—— expressed by symbolical operators, 16
—— expressed by integrals, 18
—— general properties, 19
—— repeated (coinciding) arguments, 20–21, 33, 84–86
—— case of several variables, 204
Division by omega-symbol, 178

Elderton, W. Palin, 223
Errors, generally, 2, 4, 45–48, 51–53
Error-Test, applications,
—— interpolation with divided differences, 23
—— linear interpolation, 49–50
—— numerical differentiation, 68
—— Markoff's formula for the mth difference, 70
—— formula for calculation of product-sums, 102
—— Laplace's and Gauss's summation-formulas, 114–115
—— Eulerian summation-formulas, 133, 134, 136
—— Lubbock's and Woolhouse's formulas, 151–152
Error-Test, definition, 4
Euler-Maclaurin formula,
—— for a polynomial, 122
—— general, 131
Euler's type of summation-formulas,
—— definition, 104
—— general Eulerian summation-formula, 131–132
—— first Eulerian summation-formula, 132–133, 135, 153, 192
—— second Eulerian summation-formula, 134–135, 153, 192
Euler's type of summation-formulas—continued
—— formula for repeated summation, 136–138, 202
Everett's interpolation-formulas,
—— first formula, 28, 39, 41, 44–45, 48–49, 190, 212
—— second formula, 31, 41, 51, 190, 212
Extraction of roots, 85–86

Factorial coefficients, 53–60
Factorial moments, 101
Factorials, descending, ascending, central, 8–9
Forcing-errors, 47, 51–52
Fraser, Duncan C., 14
Function-table for two variables, 203

Gaussian formulas,
—— interpolation, 26–27, 29–30
—— summation, 108–110, 114–115, 153, 190–192
—— repeated summation, 115–118, 177, 199–202
—— mechanical quadrature, 169–170
Glover, J. W., 39

Halving (interpolation to halves, to the middle),
—— one variable, 32, 75–77
—— two variables, 212
Hardy, G. F., 97
Hardy's formula, 168
Hermite's polynomials, 235–236

Indefinite sum, 90
Index law, 5
Infinite limits of summation,
—— Laplace's and Gauss's formulas, 111, 114–115
—— Euler's formulas, 132, 138
—— Lubbock's and Woolhouse's formulas, 152
Infinite power-series in theta-symbols, 184–185

Integral-formula, Dirichlet's, 93
Integration-formula, see Summation-formula.
Integration, repeated, see Repeated integration.
Interpolation-formulas, one variable,
—— divided differences, 22–25, 40, 41, 62, 64, 80
—— descending differences, 25, 26, 40–43, 186
—— ascending differences, 25, 26, 40, 41, 187
—— Gaussian, 26–27, 29–30
—— Stirling's 27–28, 31, 40, 41, 50–51, 64–66, 189
—— Bessel's, 30–32, 40, 41, 43, 45, 64, 67–68, 72, 77, 109, 189
—— Everett's first formula, 28, 39, 41, 44–45, 48–49, 190, 212
—— Everett's second formula, 31, 41, 51, 190, 212
—— central differences, 32, 187–188
—— halving of intervals, 32, 77
—— osculating interpolation, 33–34
—— Lagrange's formula, 22, 25, 33, 154, 162
Interpolation, generally, 1–2, 34–39, 52–53
Interpolation, inverse, see Inverse interpolation.
Interpolation with several variables,
—— generally, 203–204, 223
—— divided differences, 205–206, 208, 214–216, 218–219
—— descending differences, 207
—— Stirling-Stirling formula, 209–210
—— Bessel-Bessel formula, 210
—— Stirling-Bessel formula, 211
—— interpolation to the middle, 212
—— Everett's formulas, 212
—— Lagrange's formula, 216
—— three and more variables, 212–214
—— numerical illustrations, 217–223
Inverse interpolation,
—— generally, 79–80
Inverse interpolation—continued
—— by divided differences, 80–81
—— by solving a quadratic, 82
—— by repeated linear interpolation, 82–83
—— by repeated arguments, 84–86

Jacobi's theorem, 125
Jensen, J. L. W. V., 16, 184
Jensen's formula, 18
Jörgensen, N. R., 236

König, 80

Lagrange's formula,
—— one variable, 22, 25, 33, 154, 162
—— two variables, 216
Landau, 240
Laplace, 28
Laplace's summation-formula, 105, 107, 109, 114–115, 190, 192
Laplace's type of summation-formula (definition), 104
Lefrancq, E., 170
Lehmann, I., 23, 148
Leibnitz' theorem, 11, 231
Lidstone, G. J., 28, 93
Linear homogeneous difference-equation with constant coefficients, 183–184
Logarithmic-integral function, 161–162, 240
Lubbock's type of summation-formula,
—— generally, 138, 150, 153
—— first (original) formula, 139, 192
—— second formula, 143, 145, 152–153, 193–194
—— third formula, 146, 147, 194–195

Maclaurin's formula, see Taylor's formula
Markoff's formulas,
—— for the mth difference, 70, 198
—— for the mth differential coefficient, 64, 197
Mean Value, Theorem of, 3

Mechanical Cubature,
—— definition, 224
—— general formula, 226–227
—— Simpson's formula for two variables, 227, 229
—— twenty-five-term formula, 227–228
—— one-term-formula, 228
—— numerical illustrations, 228–231
Mechanical Quadrature,
—— definition, 154
—— Cotes' method, 154
—— general formulas, 157–158, 165–166
—— closed type of formulas, 158, 166
—— open type of formulas, 158, 167
—— Simpson's formula, 160–161
—— Hardy's formula, 168
—— Weddle's formula, 168–169
—— Gauss's formulas, 169–170
—— numerical illustration, 161–162
Moments, 96–97, 99–101
Moments, factorial, 101
Montel, P., 38

Narumi, S., 203
Neder, L., 203
Newton, 23, 27, 30, 105
Newton's interpolation-formula with divided differences,
—— one variable, 22–25, 40, 41, 62, 64, 80
—— two variables, 205–206, 208, 214–216, 218–219
—— three and more variables, 212–214
Numbers, $G^{(r)}_{\nu}$, 125–126; $H^{(r)}_{\nu}$, 127; $J_{2\nu}$, 117, 201; $K_{2\nu}$, 107–109, 191; L_{ν}, 105–107; $M_{2\nu}$, 109–110, 191; $N_{2\nu}$, 118, 199; $O^{[2k]}_{2m}$, 158, 159–160; $O^{[2r+1]-1}_{2m+1}$, 166; $P_{2\nu}$, 143–145, 148, 193; $Q_{2\nu}$, 145, 146–148, 195; V_{ν}, 157–158, 165–167; Λ_{ν}, 139–141; see also Bernoulli's numbers, Factorial coefficients, Differences of nothing, Central differences of nothing, Differential coefficients of nothing.
Numerical differentiation,
—— definition, 60
—— first differential coefficient, 61–62, 187–189
—— general formula, 62, 64, 65
—— Markoff's formula, 64, 197
—— central difference formulas, 66–68, 188–189, 197–198
—— second differential coefficient, 66–68
—— application of Error-Test, 68
Numerical illustrations,
—— propagation of errors in difference-table, 45–48
—— risk of neglecting remainder-term, 34–39, 52–53, 231
—— application of Error-Test, 49–50, 115, 136
—— uselessness of convergence for numerical calculations, 114
—— use of divergent series, 136
—— interpolation by descending differences, 42–43, 52
—— Bessel's interpolation-formula, 43–44
—— Stirling's interpolation-formula, 50–51
—— Everett's first formula, 44–45, 48–49
—— Everett's second formula, 51
—— construction of tables, 71, 74–79
—— inverse interpolation, 81–86
—— calculation of moments by repeated summation, 100–101
—— first Gaussian summation-formula, 112–114, 115
—— first Eulerian summation-formula, 136
—— second formula of Lubbock's type, 152–153
—— second formula of Woolhouse's type, 153
—— mechanical quadrature, 161–162

Numerical illustrations—continued
—— mechanical cubature, 228–231
—— numerical integration of differential equations, 172–176
Numerical integration, see Mechanical quadrature, Summation-formula, Differential equations.
Nyström, E. J., 177
Nörlund, N. E., 18, 119, 131, 180

Omega-symbol, 178
Open type of quadrature-formula, 158–159, 167
Osculating interpolation, 33–34

Partial summation, 89–90
Pearson, K., 41
Periodic Bernoullian function, 130
Periodic constant, 179
Picard, 241
Pincherle, S., 186
Product-sums, 101–103

Quadrature, see Mechanical quadrature, Summation-formula.

Remainder-term,
—— generally, 2, 4, 34–39, 52–53, 80, 114, 231
—— of various formulas, see these.
Repeated arguments, 20–21, 33, 84–86
Repeated integration, 103, 115–118, 136–138, 198–202
Repeated summation, 93–104, 115–118, 136–138, 198–202
Robinson, 80
Rolle's Theorem, 2–3
Roots, extraction of, 85–86
Roots of equations, see Inverse interpolation.
Runge, 18, 38, 80, 170

Sheppard, 10
Simpson's formula,
—— one variable, 160, 161
—— two variables, 227, 229
Spencer, J., 223
Steffensen, J. F., 101, 107, 118, 119
Stirling's interpolation-formula,
—— one variable, 27–28, 31, 40, 41, 50–51, 64–66, 72, 189
—— two variables, 209–211
Subdivision of intervals, see Tables, construction of.
Sum-formula, Dirichlet's, 91–92
Sum, indefinite, 90
Summation, elementary methods,
—— factorial, 87
—— polynomial, 87, 124
—— inverse factorial, 88
—— trigonometrical expressions, 89
—— generalized factorial, 91
—— generalized inverse factorial, 91
—— partial summation, 89–90
Summation-formula,
—— Laplace's type, generally, 104
—— Laplace's, 105, 107, 109, 114–115, 190, 192
—— Gaussian, first, 108, 114–117, 177, 190–192, 200–202
—— Gaussian, second, 109–110, 114–115, 117–118, 177, 191–192, 199–200
—— Euler's type, generally, 104, 153, 154
—— Eulerian, general, 131–132
—— Eulerian, first, 132–133, 135–138, 192, 202
—— Eulerian, second, 134–135, 192
—— Lubbock's type, generally, 138, 150, 153
—— Lubbock's formula, 139, 192
—— second formula of Lubbock's type, 143, 145, 152–153, 193–194
—— third formula of Lubbock's type, 146, 147, 194–195
—— Woolhouse's type, generally, 138, 153
—— Woolhouse's formula, 148–150, 195–196
—— second formula of Woolhouse's type, 150–151, 153, 195–196

Summation, repeated, see Repeated summation.
Sum-table,
—— for summation from the top, 94
—— for summation from the bottom, 98
Symbols, calculus of, 178–202
Symbols, E, 4–5; Δ, ∇, δ, 7; D, 8; $\square$, 10; θ_p, 16; Σ, 93; S, 95; Σ', S', 97; Ω, Ω^{-1}, 178; Δ^{-1}, ∇^{-1}, δ^{-1}, D^{-1}, 178–179; Θ, 184

Tables, construction of, 71–79
Tables of various numbers, see these.
Taylor's formula,
—— one variable, 25–26, 186
—— two variables, 207, 216
Tchebychef, 236
Theta-symbol, 184
Thiele, 10, 19, 47, 72, 84
Thompson, A. J., 39

Walther, A., 155
Watson, 3, 240
Weddle's formula, 168–169
Whittaker, 3, 80, 240
Willers, 18, 170
Woolhouse's type of summation-formula,
—— generally, 138, 153
—— Woolhouse's formula, 148–150, 195–196
—— second formula of Woolhouse's type, 150–151, 153, 195–196

A CATALOG OF SELECTED

DOVER BOOKS

IN SCIENCE AND MATHEMATICS

Astronomy

BURNHAM'S CELESTIAL HANDBOOK, Robert Burnham, Jr. Thorough guide to the stars beyond our solar system. Exhaustive treatment. Alphabetical by constellation: Andromeda to Cetus in Vol. 1; Chamaeleon to Orion in Vol. 2; and Pavo to Vulpecula in Vol. 3. Hundreds of illustrations. Index in Vol. 3. 2,000pp. 6⅛ x 9¼.
Vol. I: 0-486-23567-X
Vol. II: 0-486-23568-8
Vol. III: 0-486-23673-0

EXPLORING THE MOON THROUGH BINOCULARS AND SMALL TELESCOPES, Ernest H. Cherrington, Jr. Informative, profusely illustrated guide to locating and identifying craters, rills, seas, mountains, other lunar features. Newly revised and updated with special section of new photos. Over 100 photos and diagrams. 240pp. 8¼ x 11. 0-486-24491-1

THE EXTRATERRESTRIAL LIFE DEBATE, 1750–1900, Michael J. Crowe. First detailed, scholarly study in English of the many ideas that developed from 1750 to 1900 regarding the existence of intelligent extraterrestrial life. Examines ideas of Kant, Herschel, Voltaire, Percival Lowell, many other scientists and thinkers. 16 illustrations. 704pp. 5⅜ x 8½. 0-486-40675-X

THEORIES OF THE WORLD FROM ANTIQUITY TO THE COPERNICAN REVOLUTION, Michael J. Crowe. Newly revised edition of an accessible, enlightening book recreates the change from an earth-centered to a sun-centered conception of the solar system. 242pp. 5⅜ x 8½. 0-486-41444-2

A HISTORY OF ASTRONOMY, A. Pannekoek. Well-balanced, carefully reasoned study covers such topics as Ptolemaic theory, work of Copernicus, Kepler, Newton, Eddington's work on stars, much more. Illustrated. References. 521pp. 5⅜ x 8½.
0-486-65994-1

A COMPLETE MANUAL OF AMATEUR ASTRONOMY: TOOLS AND TECHNIQUES FOR ASTRONOMICAL OBSERVATIONS, P. Clay Sherrod with Thomas L. Koed. Concise, highly readable book discusses: selecting, setting up and maintaining a telescope; amateur studies of the sun; lunar topography and occultations; observations of Mars, Jupiter, Saturn, the minor planets and the stars; an introduction to photoelectric photometry; more. 1981 ed. 124 figures. 25 halftones. 37 tables. 335pp. 6½ x 9¼. 0-486-40675-X

AMATEUR ASTRONOMER'S HANDBOOK, J. B. Sidgwick. Timeless, comprehensive coverage of telescopes, mirrors, lenses, mountings, telescope drives, micrometers, spectroscopes, more. 189 illustrations. 576pp. 5⅝ x 8¼. (Available in U.S. only.)
0-486-24034-7

STARS AND RELATIVITY, Ya. B. Zel'dovich and I. D. Novikov. Vol. 1 of *Relativistic Astrophysics* by famed Russian scientists. General relativity, properties of matter under astrophysical conditions, stars, and stellar systems. Deep physical insights, clear presentation. 1971 edition. References. 544pp. 5⅝ x 8¼. 0-486-69424-0

Chemistry

THE SCEPTICAL CHYMIST: THE CLASSIC 1661 TEXT, Robert Boyle. Boyle defines the term "element," asserting that all natural phenomena can be explained by the motion and organization of primary particles. 1911 ed. viii+232pp. 5⅜ x 8½.
0-486-42825-7

RADIOACTIVE SUBSTANCES, Marie Curie. Here is the celebrated scientist's doctoral thesis, the prelude to her receipt of the 1903 Nobel Prize. Curie discusses establishing atomic character of radioactivity found in compounds of uranium and thorium; extraction from pitchblende of polonium and radium; isolation of pure radium chloride; determination of atomic weight of radium; plus electric, photographic, luminous, heat, color effects of radioactivity. ii+94pp. 5⅜ x 8½. 0-486-42550-9

CHEMICAL MAGIC, Leonard A. Ford. Second Edition, Revised by E. Winston Grundmeier. Over 100 unusual stunts demonstrating cold fire, dust explosions, much more. Text explains scientific principles and stresses safety precautions. 128pp. 5⅜ x 8½. 0-486-67628-5

THE DEVELOPMENT OF MODERN CHEMISTRY, Aaron J. Ihde. Authoritative history of chemistry from ancient Greek theory to 20th-century innovation. Covers major chemists and their discoveries. 209 illustrations. 14 tables. Bibliographies. Indices. Appendices. 851pp. 5⅜ x 8½. 0-486-64235-6

CATALYSIS IN CHEMISTRY AND ENZYMOLOGY, William P. Jencks. Exceptionally clear coverage of mechanisms for catalysis, forces in aqueous solution, carbonyl- and acyl-group reactions, practical kinetics, more. 864pp. 5⅜ x 8½.
0-486-65460-5

ELEMENTS OF CHEMISTRY, Antoine Lavoisier. Monumental classic by founder of modern chemistry in remarkable reprint of rare 1790 Kerr translation. A must for every student of chemistry or the history of science. 539pp. 5⅜ x 8½. 0-486-64624-6

THE HISTORICAL BACKGROUND OF CHEMISTRY, Henry M. Leicester. Evolution of ideas, not individual biography. Concentrates on formulation of a coherent set of chemical laws. 260pp. 5⅜ x 8½. 0-486-61053-5

A SHORT HISTORY OF CHEMISTRY, J. R. Partington. Classic exposition explores origins of chemistry, alchemy, early medical chemistry, nature of atmosphere, theory of valency, laws and structure of atomic theory, much more. 428pp. 5⅜ x 8½. (Available in U.S. only.) 0-486-65977-1

GENERAL CHEMISTRY, Linus Pauling. Revised 3rd edition of classic first-year text by Nobel laureate. Atomic and molecular structure, quantum mechanics, statistical mechanics, thermodynamics correlated with descriptive chemistry. Problems. 992pp. 5⅜ x 8½. 0-486-65622-5

FROM ALCHEMY TO CHEMISTRY, John Read. Broad, humanistic treatment focuses on great figures of chemistry and ideas that revolutionized the science. 50 illustrations. 240pp. 5⅜ x 8½. 0-486-28690-8

Engineering

DE RE METALLICA, Georgius Agricola. The famous Hoover translation of greatest treatise on technological chemistry, engineering, geology, mining of early modern times (1556). All 289 original woodcuts. 638pp. 6¾ x 11. 0-486-60006-8

FUNDAMENTALS OF ASTRODYNAMICS, Roger Bate et al. Modern approach developed by U.S. Air Force Academy. Designed as a first course. Problems, exercises. Numerous illustrations. 455pp. 5⅜ x 8½. 0-486-60061-0

DYNAMICS OF FLUIDS IN POROUS MEDIA, Jacob Bear. For advanced students of ground water hydrology, soil mechanics and physics, drainage and irrigation engineering and more. 335 illustrations. Exercises, with answers. 784pp. 6⅛ x 9¼. 0-486-65675-6

THEORY OF VISCOELASTICITY (Second Edition), Richard M. Christensen. Complete consistent description of the linear theory of the viscoelastic behavior of materials. Problem-solving techniques discussed. 1982 edition. 29 figures. xiv+364pp. 6⅛ x 9¼. 0-486-42880-X

MECHANICS, J. P. Den Hartog. A classic introductory text or refresher. Hundreds of applications and design problems illuminate fundamentals of trusses, loaded beams and cables, etc. 334 answered problems. 462pp. 5⅜ x 8½. 0-486-60754-2

MECHANICAL VIBRATIONS, J. P. Den Hartog. Classic textbook offers lucid explanations and illustrative models, applying theories of vibrations to a variety of practical industrial engineering problems. Numerous figures. 233 problems, solutions. Appendix. Index. Preface. 436pp. 5⅜ x 8½. 0-486-64785-4

STRENGTH OF MATERIALS, J. P. Den Hartog. Full, clear treatment of basic material (tension, torsion, bending, etc.) plus advanced material on engineering methods, applications. 350 answered problems. 323pp. 5⅜ x 8½. 0-486-60755-0

A HISTORY OF MECHANICS, René Dugas. Monumental study of mechanical principles from antiquity to quantum mechanics. Contributions of ancient Greeks, Galileo, Leonardo, Kepler, Lagrange, many others. 671pp. 5⅜ x 8½. 0-486-65632-2

STABILITY THEORY AND ITS APPLICATIONS TO STRUCTURAL MECHANICS, Clive L. Dym. Self-contained text focuses on Koiter postbuckling analyses, with mathematical notions of stability of motion. Basing minimum energy principles for static stability upon dynamic concepts of stability of motion, it develops asymptotic buckling and postbuckling analyses from potential energy considerations, with applications to columns, plates, and arches. 1974 ed. 208pp. 5⅜ x 8½. 0-486-42541-X

METAL FATIGUE, N. E. Frost, K. J. Marsh, and L. P. Pook. Definitive, clearly written, and well-illustrated volume addresses all aspects of the subject, from the historical development of understanding metal fatigue to vital concepts of the cyclic stress that causes a crack to grow. Includes 7 appendixes. 544pp. 5⅜ x 8½. 0-486-40927-9

ROCKETS, Robert Goddard. Two of the most significant publications in the history of rocketry and jet propulsion: "A Method of Reaching Extreme Altitudes" (1919) and "Liquid Propellant Rocket Development" (1936). 128pp. 5⅜ x 8½. 0-486-42537-1

STATISTICAL MECHANICS: PRINCIPLES AND APPLICATIONS, Terrell L. Hill. Standard text covers fundamentals of statistical mechanics, applications to fluctuation theory, imperfect gases, distribution functions, more. 448pp. 5⅜ x 8½. 0-486-65390-0

ENGINEERING AND TECHNOLOGY 1650–1750: ILLUSTRATIONS AND TEXTS FROM ORIGINAL SOURCES, Martin Jensen. Highly readable text with more than 200 contemporary drawings and detailed engravings of engineering projects dealing with surveying, leveling, materials, hand tools, lifting equipment, transport and erection, piling, bailing, water supply, hydraulic engineering, and more. Among the specific projects outlined-transporting a 50-ton stone to the Louvre, erecting an obelisk, building timber locks, and dredging canals. 207pp. 8⅜ x 11¼. 0-486-42232-1

THE VARIATIONAL PRINCIPLES OF MECHANICS, Cornelius Lanczos. Graduate level coverage of calculus of variations, equations of motion, relativistic mechanics, more. First inexpensive paperbound edition of classic treatise. Index. Bibliography. 418pp. 5⅜ x 8½. 0-486-65067-7

PROTECTION OF ELECTRONIC CIRCUITS FROM OVERVOLTAGES, Ronald B. Standler. Five-part treatment presents practical rules and strategies for circuits designed to protect electronic systems from damage by transient overvoltages. 1989 ed. xxiv+434pp. 6⅛ x 9¼. 0-486-42552-5

ROTARY WING AERODYNAMICS, W. Z. Stepniewski. Clear, concise text covers aerodynamic phenomena of the rotor and offers guidelines for helicopter performance evaluation. Originally prepared for NASA. 537 figures. 640pp. 6⅛ x 9¼. 0-486-64647-5

INTRODUCTION TO SPACE DYNAMICS, William Tyrrell Thomson. Comprehensive, classic introduction to space-flight engineering for advanced undergraduate and graduate students. Includes vector algebra, kinematics, transformation of coordinates. Bibliography. Index. 352pp. 5⅜ x 8½. 0-486-65113-4

HISTORY OF STRENGTH OF MATERIALS, Stephen P. Timoshenko. Excellent historical survey of the strength of materials with many references to the theories of elasticity and structure. 245 figures. 452pp. 5⅜ x 8½. 0-486-61187-6

ANALYTICAL FRACTURE MECHANICS, David J. Unger. Self-contained text supplements standard fracture mechanics texts by focusing on analytical methods for determining crack-tip stress and strain fields. 336pp. 6⅛ x 9¼. 0-486-41737-9

STATISTICAL MECHANICS OF ELASTICITY, J. H. Weiner. Advanced, self-contained treatment illustrates general principles and elastic behavior of solids. Part 1, based on classical mechanics, studies thermoelastic behavior of crystalline and polymeric solids. Part 2, based on quantum mechanics, focuses on interatomic force laws, behavior of solids, and thermally activated processes. For students of physics and chemistry and for polymer physicists. 1983 ed. 96 figures. 496pp. 5⅜ x 8½. 0-486-42260-7

Mathematics

FUNCTIONAL ANALYSIS (Second Corrected Edition), George Bachman and Lawrence Narici. Excellent treatment of subject geared toward students with background in linear algebra, advanced calculus, physics and engineering. Text covers introduction to inner-product spaces, normed, metric spaces, and topological spaces; complete orthonormal sets, the Hahn-Banach Theorem and its consequences, and many other related subjects. 1966 ed. 544pp. 6⅛ x 9¼. 0-486-40251-7

ASYMPTOTIC EXPANSIONS OF INTEGRALS, Norman Bleistein & Richard A. Handelsman. Best introduction to important field with applications in a variety of scientific disciplines. New preface. Problems. Diagrams. Tables. Bibliography. Index. 448pp. 5⅜ x 8½. 0-486-65082-0

VECTOR AND TENSOR ANALYSIS WITH APPLICATIONS, A. I. Borisenko and I. E. Tarapov. Concise introduction. Worked-out problems, solutions, exercises. 257pp. 5⅝ x 8¼. 0-486-63833-2

AN INTRODUCTION TO ORDINARY DIFFERENTIAL EQUATIONS, Earl A. Coddington. A thorough and systematic first course in elementary differential equations for undergraduates in mathematics and science, with many exercises and problems (with answers). Index. 304pp. 5⅜ x 8½. 0-486-65942-9

FOURIER SERIES AND ORTHOGONAL FUNCTIONS, Harry F. Davis. An incisive text combining theory and practical example to introduce Fourier series, orthogonal functions and applications of the Fourier method to boundary-value problems. 570 exercises. Answers and notes. 416pp. 5⅜ x 8½. 0-486-65973-9

COMPUTABILITY AND UNSOLVABILITY, Martin Davis. Classic graduate-level introduction to theory of computability, usually referred to as theory of recurrent functions. New preface and appendix. 288pp. 5⅜ x 8½. 0-486-61471-9

ASYMPTOTIC METHODS IN ANALYSIS, N. G. de Bruijn. An inexpensive, comprehensive guide to asymptotic methods–the pioneering work that teaches by explaining worked examples in detail. Index. 224pp. 5⅜ x 8½ 0-486-64221-6

APPLIED COMPLEX VARIABLES, John W. Dettman. Step-by-step coverage of fundamentals of analytic function theory–plus lucid exposition of five important applications: Potential Theory; Ordinary Differential Equations; Fourier Transforms; Laplace Transforms; Asymptotic Expansions. 66 figures. Exercises at chapter ends. 512pp. 5⅜ x 8½. 0-486-64670-X

INTRODUCTION TO LINEAR ALGEBRA AND DIFFERENTIAL EQUATIONS, John W. Dettman. Excellent text covers complex numbers, determinants, orthonormal bases, Laplace transforms, much more. Exercises with solutions. Undergraduate level. 416pp. 5⅜ x 8½. 0-486-65191-6

RIEMANN'S ZETA FUNCTION, H. M. Edwards. Superb, high-level study of landmark 1859 publication entitled "On the Number of Primes Less Than a Given Magnitude" traces developments in mathematical theory that it inspired. xiv+315pp. 5⅜ x 8½. 0-486-41740-9

Math–Decision Theory, Statistics, Probability

ELEMENTARY DECISION THEORY, Herman Chernoff and Lincoln E. Moses. Clear introduction to statistics and statistical theory covers data processing, probability and random variables, testing hypotheses, much more. Exercises. 364pp. 5⅜ x 8½. 0-486-65218-1

STATISTICS MANUAL, Edwin L. Crow et al. Comprehensive, practical collection of classical and modern methods prepared by U.S. Naval Ordnance Test Station. Stress on use. Basics of statistics assumed. 288pp. 5⅜ x 8½. 0-486-60599-X

SOME THEORY OF SAMPLING, William Edwards Deming. Analysis of the problems, theory and design of sampling techniques for social scientists, industrial managers and others who find statistics important at work. 61 tables. 90 figures. xvii +602pp. 5⅜ x 8½. 0-486-64684-X

LINEAR PROGRAMMING AND ECONOMIC ANALYSIS, Robert Dorfman, Paul A. Samuelson and Robert M. Solow. First comprehensive treatment of linear programming in standard economic analysis. Game theory, modern welfare economics, Leontief input-output, more. 525pp. 5⅜ x 8½. 0-486-65491-5

PROBABILITY: AN INTRODUCTION, Samuel Goldberg. Excellent basic text covers set theory, probability theory for finite sample spaces, binomial theorem, much more. 360 problems. Bibliographies. 322pp. 5⅜ x 8½. 0-486-65252-1

GAMES AND DECISIONS: INTRODUCTION AND CRITICAL SURVEY, R. Duncan Luce and Howard Raiffa. Superb nontechnical introduction to game theory, primarily applied to social sciences. Utility theory, zero-sum games, n-person games, decision-making, much more. Bibliography. 509pp. 5⅜ x 8½. 0-486-65943-7

INTRODUCTION TO THE THEORY OF GAMES, J. C. C. McKinsey. This comprehensive overview of the mathematical theory of games illustrates applications to situations involving conflicts of interest, including economic, social, political, and military contexts. Appropriate for advanced undergraduate and graduate courses; advanced calculus a prerequisite. 1952 ed. x+372pp. 5⅜ x 8½. 0-486-42811-7

FIFTY CHALLENGING PROBLEMS IN PROBABILITY WITH SOLUTIONS, Frederick Mosteller. Remarkable puzzlers, graded in difficulty, illustrate elementary and advanced aspects of probability. Detailed solutions. 88pp. 5⅜ x 8½. 65355-2

PROBABILITY THEORY: A CONCISE COURSE, Y. A. Rozanov. Highly readable, self-contained introduction covers combination of events, dependent events, Bernoulli trials, etc. 148pp. 5⅜ x 8¼. 0-486-63544-9

STATISTICAL METHOD FROM THE VIEWPOINT OF QUALITY CONTROL, Walter A. Shewhart. Important text explains regulation of variables, uses of statistical control to achieve quality control in industry, agriculture, other areas. 192pp. 5⅜ x 8½. 0-486-65232-7

Math–Geometry and Topology

ELEMENTARY CONCEPTS OF TOPOLOGY, Paul Alexandroff. Elegant, intuitive approach to topology from set-theoretic topology to Betti groups; how concepts of topology are useful in math and physics. 25 figures. 57pp. 5⅜ x 8½. 0-486-60747-X

COMBINATORIAL TOPOLOGY, P. S. Alexandrov. Clearly written, well-organized, three-part text begins by dealing with certain classic problems without using the formal techniques of homology theory and advances to the central concept, the Betti groups. Numerous detailed examples. 654pp. 5⅜ x 8½. 0-486-40179-0

EXPERIMENTS IN TOPOLOGY, Stephen Barr. Classic, lively explanation of one of the byways of mathematics. Klein bottles, Moebius strips, projective planes, map coloring, problem of the Koenigsberg bridges, much more, described with clarity and wit. 43 figures. 210pp. 5⅜ x 8½. 0-486-25933-1

THE GEOMETRY OF RENÉ DESCARTES, René Descartes. The great work founded analytical geometry. Original French text, Descartes's own diagrams, together with definitive Smith-Latham translation. 244pp. 5⅜ x 8½. 0-486-60068-8

EUCLIDEAN GEOMETRY AND TRANSFORMATIONS, Clayton W. Dodge. This introduction to Euclidean geometry emphasizes transformations, particularly isometries and similarities. Suitable for undergraduate courses, it includes numerous examples, many with detailed answers. 1972 ed. viii+296pp. 6⅛ x 9¼. 0-486-43476-1

PRACTICAL CONIC SECTIONS: THE GEOMETRIC PROPERTIES OF ELLIPSES, PARABOLAS AND HYPERBOLAS, J. W. Downs. This text shows how to create ellipses, parabolas, and hyperbolas. It also presents historical background on their ancient origins and describes the reflective properties and roles of curves in design applications. 1993 ed. 98 figures. xii+100pp. 6½ x 9¼. 0-486-42876-1

THE THIRTEEN BOOKS OF EUCLID'S ELEMENTS, translated with introduction and commentary by Sir Thomas L. Heath. Definitive edition. Textual and linguistic notes, mathematical analysis. 2,500 years of critical commentary. Unabridged. 1,414pp. 5⅜ x 8½. Three-vol. set.
Vol. I: 0-486-60088-2 Vol. II: 0-486-60089-0 Vol. III: 0-486-60090-4

SPACE AND GEOMETRY: IN THE LIGHT OF PHYSIOLOGICAL, PSYCHOLOGICAL AND PHYSICAL INQUIRY, Ernst Mach. Three essays by an eminent philosopher and scientist explore the nature, origin, and development of our concepts of space, with a distinctness and precision suitable for undergraduate students and other readers. 1906 ed. vi+148pp. 5⅜ x 8½. 0-486-43909-7

GEOMETRY OF COMPLEX NUMBERS, Hans Schwerdtfeger. Illuminating, widely praised book on analytic geometry of circles, the Moebius transformation, and two-dimensional non-Euclidean geometries. 200pp. 5⅝ x 8¼. 0-486-63830-8

DIFFERENTIAL GEOMETRY, Heinrich W. Guggenheimer. Local differential geometry as an application of advanced calculus and linear algebra. Curvature, transformation groups, surfaces, more. Exercises. 62 figures. 378pp. 5⅜ x 8½. 0-486-63433-7

History of Math

THE WORKS OF ARCHIMEDES, Archimedes (T. L. Heath, ed.). Topics include the famous problems of the ratio of the areas of a cylinder and an inscribed sphere; the measurement of a circle; the properties of conoids, spheroids, and spirals; and the quadrature of the parabola. Informative introduction. clxxxvi+326pp. 5⅜ x 8½. 0-486-42084-1

A SHORT ACCOUNT OF THE HISTORY OF MATHEMATICS, W. W. Rouse Ball. One of clearest, most authoritative surveys from the Egyptians and Phoenicians through 19th-century figures such as Grassman, Galois, Riemann. Fourth edition. 522pp. 5⅜ x 8½. 0-486-20630-0

THE HISTORY OF THE CALCULUS AND ITS CONCEPTUAL DEVELOPMENT, Carl B. Boyer. Origins in antiquity, medieval contributions, work of Newton, Leibniz, rigorous formulation. Treatment is verbal. 346pp. 5⅜ x 8½. 0-486-60509-4

THE HISTORICAL ROOTS OF ELEMENTARY MATHEMATICS, Lucas N. H. Bunt, Phillip S. Jones, and Jack D. Bedient. Fundamental underpinnings of modern arithmetic, algebra, geometry and number systems derived from ancient civilizations. 320pp. 5⅜ x 8½. 0-486-25563-8

A HISTORY OF MATHEMATICAL NOTATIONS, Florian Cajori. This classic study notes the first appearance of a mathematical symbol and its origin, the competition it encountered, its spread among writers in different countries, its rise to popularity, its eventual decline or ultimate survival. Original 1929 two-volume edition presented here in one volume. xxviii+820pp. 5⅜ x 8½. 0-486-67766-4

GAMES, GODS & GAMBLING: A HISTORY OF PROBABILITY AND STATISTICAL IDEAS, F. N. David. Episodes from the lives of Galileo, Fermat, Pascal, and others illustrate this fascinating account of the roots of mathematics. Features thought-provoking references to classics, archaeology, biography, poetry. 1962 edition. 304pp. 5⅜ x 8½. (Available in U.S. only.) 0-486-40023-9

OF MEN AND NUMBERS: THE STORY OF THE GREAT MATHEMATICIANS, Jane Muir. Fascinating accounts of the lives and accomplishments of history's greatest mathematical minds–Pythagoras, Descartes, Euler, Pascal, Cantor, many more. Anecdotal, illuminating. 30 diagrams. Bibliography. 256pp. 5⅜ x 8½. 0-486-28973-7

HISTORY OF MATHEMATICS, David E. Smith. Nontechnical survey from ancient Greece and Orient to late 19th century; evolution of arithmetic, geometry, trigonometry, calculating devices, algebra, the calculus. 362 illustrations. 1,355pp. 5⅜ x 8½. Two-vol. set. Vol. I: 0-486-20429-4 Vol. II: 0-486-20430-8

A CONCISE HISTORY OF MATHEMATICS, Dirk J. Struik. The best brief history of mathematics. Stresses origins and covers every major figure from ancient Near East to 19th century. 41 illustrations. 195pp. 5⅜ x 8½. 0-486-60255-9

Physics

OPTICAL RESONANCE AND TWO-LEVEL ATOMS, L. Allen and J. H. Eberly. Clear, comprehensive introduction to basic principles behind all quantum optical resonance phenomena. 53 illustrations. Preface. Index. 256pp. 5⅜ x 8½. 0-486-65533-4

QUANTUM THEORY, David Bohm. This advanced undergraduate-level text presents the quantum theory in terms of qualitative and imaginative concepts, followed by specific applications worked out in mathematical detail. Preface. Index. 655pp. 5⅜ x 8½. 0-486-65969-0

ATOMIC PHYSICS (8th EDITION), Max Born. Nobel laureate's lucid treatment of kinetic theory of gases, elementary particles, nuclear atom, wave-corpuscles, atomic structure and spectral lines, much more. Over 40 appendices, bibliography. 495pp. 5⅜ x 8½. 0-486-65984-4

A SOPHISTICATE'S PRIMER OF RELATIVITY, P. W. Bridgman. Geared toward readers already acquainted with special relativity, this book transcends the view of theory as a working tool to answer natural questions: What is a frame of reference? What is a "law of nature"? What is the role of the "observer"? Extensive treatment, written in terms accessible to those without a scientific background. 1983 ed. xlviii+172pp. 5⅜ x 8½. 0-486-42549-5

AN INTRODUCTION TO HAMILTONIAN OPTICS, H. A. Buchdahl. Detailed account of the Hamiltonian treatment of aberration theory in geometrical optics. Many classes of optical systems defined in terms of the symmetries they possess. Problems with detailed solutions. 1970 edition. xv + 360pp. 5⅜ x 8½. 0-486-67597-1

PRIMER OF QUANTUM MECHANICS, Marvin Chester. Introductory text examines the classical quantum bead on a track: its state and representations; operator eigenvalues; harmonic oscillator and bound bead in a symmetric force field; and bead in a spherical shell. Other topics include spin, matrices, and the structure of quantum mechanics; the simplest atom; indistinguishable particles; and stationary-state perturbation theory. 1992 ed. xiv+314pp. 6⅛ x 9¼. 0-486-42878-8

LECTURES ON QUANTUM MECHANICS, Paul A. M. Dirac. Four concise, brilliant lectures on mathematical methods in quantum mechanics from Nobel Prize-winning quantum pioneer build on idea of visualizing quantum theory through the use of classical mechanics. 96pp. 5⅜ x 8½. 0-486-41713-1

THIRTY YEARS THAT SHOOK PHYSICS: THE STORY OF QUANTUM THEORY, George Gamow. Lucid, accessible introduction to influential theory of energy and matter. Careful explanations of Dirac's anti-particles, Bohr's model of the atom, much more. 12 plates. Numerous drawings. 240pp. 5⅜ x 8½. 0-486-24895-X

ELECTRONIC STRUCTURE AND THE PROPERTIES OF SOLIDS: THE PHYSICS OF THE CHEMICAL BOND, Walter A. Harrison. Innovative text offers basic understanding of the electronic structure of covalent and ionic solids, simple metals, transition metals and their compounds. Problems. 1980 edition. 582pp. 6⅛ x 9¼. 0-486-66021-4